THE REALITY REVOLUTION

THE REALITY REVOLUTION

THE MIND-BLOWING MOVEMENT
TO HACK YOUR REALITY

BRIAN SCOTT

COPYRIGHT © 2020 BRIAN SCOTT
All rights reserved.

THE REALITY REVOLUTION
The Mind-Blowing Movement to Hack Your Reality

ISBN 978-1-5445-0620-3 *Hardcover*
       978-1-5445-0618-0 *Paperback*
       978-1-5445-0619-7 *Ebook*
       978-1-5445-0621-0 *Audiobook*

CONTENTS

INTRODUCTION 9

PART I: REALITY ISN'T WHAT YOU THINK IT IS
1. QUANTUM MECHANICS 27
2. BLOCKS TO REALITY CREATION 55
3. PLAYING WITH REALITY 87

PART II: TAKE CONTROL OF YOUR REALITY
4. HACK YOUR SUBCONSCIOUS 113
5. WHY ENERGY MATTERS 143
6. TUNE IN TO YOUR INTUITION 177
7. DEFINE YOUR DREAM LIFE 211

PART III: A NEW REVOLUTION
8. THE HEALTH REVOLUTION 235
9. THE PROSPERITY REVOLUTION 271
10. THE LOVE REVOLUTION 307

PART IV: HACKING YOUR REALITY
11. HOW DO WE CONSCIOUSLY SURF THROUGH PARALLEL REALITIES? 333
12. KEY CONCEPTS TO UNDERSTAND 365
13. MIND HACKS TO CHANGE YOUR REALITY 375

CONCLUSION 391
ACKNOWLEDGMENTS 397
ABOUT THE AUTHOR 399

Warning: This is not your normal book. What follows could drastically change your thoughts, beliefs, and life. Proceed with caution.

INTRODUCTION

"I know I was born and I know that I'll die. The in between is mine."

—EDDIE VEDDER

The night my home team won the Superbowl (go Broncos!), someone broke into my home and shot me. Or rather, a version of them shot a version of me.

I'll explain.

The logistics of my story are not uncommon. It was an attempted burglary gone wrong. I heard scratching at the back door and thought it was my cat, but when I opened it, right in front of me stood a kid in a hoodie. He was in the act of breaking into my home, and held a gun pointed directly at me. He fired a shot. I turned and ran back into the house, called the cops, and the whole thing ended with a bullet bouncing off my back. I'm lucky to be alive.

As anyone who has lived through a traumatic event

knows, no matter how straightforward the facts are, the experience itself is never simple and the details rarely feel commonplace. But I believe there's even more to it than that. In fact, I believe that each moment in time is much more than it appears to be. To explain, I want you to first understand what actually happened that night.

When I went to the door, still riding that post-Superbowl high, I was just worried about my cat. My cat had only ever been indoors, and one of my nightmares is that I will accidentally let him out and into danger. My fenced backyard has a pool, and beyond that is a park. So, when I heard an odd scratching at the sliding back door—one of those doors that can't be opened or closed without a good amount of leverage—I didn't think twice about the voice in my head telling me, *Get up. You need to get up.* I figured it was just the voice of concern for my cat speaking, and in the excitement of the evening, the cat must have somehow gotten out.

At the prompting of that inner voice, I shuffled over to the door, still half-listening to the recaps and news broadcasts singing the Broncos' praise. When I looked up, I was stunned to see there was a gun in my face.

Now the voice told me, *Shut the door*, and again I listened. Everything seemed to slow down from there. I heaved the doubled-paned glass door shut and turned to run. A *pop* sounded as the .22 pistol unleashed a bullet that shattered both layers of glass. I could hear each layer of glass crack in slow motion. I let go and found a neutral place inside of me, as if I was observing it from afar. I ran without stopping to think about what direction I should run in. Somehow, I already knew.

I felt something bump against my back as I ran to my bedroom, and when I got there, I realized that I wasn't alone. I was in the middle of a home invasion robbery, and someone else had come in through my bedroom.

He started shooting at me. I could feel the heat and see a flash of metal as a bullet crossed in front of my eyes and hit the wall, then three or four more flew behind me as I ran once again, this time toward the garage. From there, I was able to hide and call the police, worried all the while that the intruders would find me hiding in the garage.

When they arrived, the police saw blood on my back and immediately sent me to the hospital. There, I found out that the bullet that had gone through two panes of glass, found its way to my back, *and bounced off of it*. It was a miracle.

I had been given a second chance. I thought about what that might mean. I thought about my kids, the life I had led up to that point, and whether or not I had done anything to deserve this second chance.

When I peel back the layers of that night and all of the thoughts that were running through my head simultaneously, I now realize that something else was happening too. The thing that was happening wasn't just adrenaline and fear or gratitude and a sense of mortality. It felt like

I was living through a memory. It felt like I had already gone through all of this before, and I could see different versions of myself in those moments, each with different outcomes. I'm not talking about running through potential scenarios—it felt like I was actually *remembering* those experiences that hadn't happened and seeing them play out.

I could see myself lying in a pool of blood. I could see myself crawling over to the kitchen table and being hit by another bullet. I don't know any other way to describe this than by explaining to you that none of this was the result of fear; it was experience. I intricately knew about all the possible realities that could play out before me as if they were real. All the versions of me looked real, as if I could walk over and touch them.

WHAT HAPPENED AFTER THE JUMP

The ambulance took me to the hospital in the same state I had been in when I was watching TV—in my boxers—so I had no wallet, no ID, and no clothes with me at the hospital. Since I didn't have any serious injuries, they discharged me that same night. I called an Uber and went home in slippers and a paper gown from the hospital.

When the Uber dropped me off, I realized that the cul-de-sac where I lived had been cordoned off. Helicopters flew overhead, cops held K-9s at the end of harnesses, and a standoff was in progress. I started toward my driveway, only to be stopped at the barricade. The police told me there was a hostage situation in place. They had traced the culprits back to a house across the street, where the whole ruined escapade was ending in a shambles. Standing there half-naked on the wrong side of the yellow caution tape, without any way to prove who I was, I just wanted to get back into my house and check on my cat (luckily, he decided to stick around!).

When the dust finally settled and everything was back in place, I realized that nothing was back in place at all. Things just seemed...off. At first, I blew it off. I thought I might be experiencing some kind of post-traumatic stress disorder that made everything seem a little strange—but enough oddities and reality shifts started to pile up that I could no longer ignore them.

Lots of stuff had changed. My kids were different. I noticed a lamp in the corner of the room that I did not remember buying. People called me acting as if we had talked just a few days before, when I had not spoken to them in years. There was suddenly an old building on the corner where there had previously been an empty lot. These are only some of the numerous shifts in my reality that were impossible to ignore.

Something had happened, though. A reality shift had happened. Previous to this incident I had been toying with and hoping to discover reality shifts, but I had no real concept of what they would actually mean.

I began to think that my near-death experience had

been enough to propel me into a significantly different reality, but my scientific mind needed proof.

Had I written that sentence fifty or even twenty years ago, you would probably think I was insane. Perhaps even today you might wonder if I'm a little "off" (although, if that were the case, you probably wouldn't be reading this book).

Conversations about parallel realities are becoming more and more commonplace as people notice little shifts in their memories. Things that seem to be concrete memories for millions of people are supposedly not true. I'm talking about seemingly little details like this. Despite what you know you remember, it is "fact" that Captain Crunch cereal has always been Cap'n Crunch, Jiffy peanut butter has always been Jif, and Ed McMahon never worked for Publisher's Clearinghouse. There's even a name for instances like this when you *know* you remember something, only to find out that it is incorrect or never happened; it's called the Mandela Effect.

By the time I noticed that the world around me had shifted, I had already done some studying about the Mandela Effect but, still, I didn't recognize it immediately. I had even embraced the ideas of surfing between different realities—not because I was out of my mind, but because I had recently become much more present in my own mind.

EXPERIMENTING WITH REALITY

As a person who reads voraciously, listens to audiobooks at four times the normal speed so that I can digest more information more quickly, and accumulates information endlessly, I thought I had a basic grasp of the laws of the universe. At least I thought that until the night I lost everything. And, no, I didn't lose it all on the night of the home invasion, but on the night the mother of my kids left me and moved from our home in Colorado to Oregon. I was devastated.

A good friend of mine in California had a home for rent that he offered me. It was in this home that the invasion occurred. I moved there so that I could have a fresh start, but instead I slipped into the abyss. My addictive personality landed on alcohol. I started drinking and could not stop. The pool in the backyard became a focus for my suicidal ideation. I had flashes in my mind of weighing myself down and sinking to the bottom. My kids were the only thing that held me back—I did not want them to find out I killed myself. My wonderful sister (who had found happiness after experiencing the unimaginable accidental death of her husband) advised me to take responsibility for everything that had happened in my life and to change my thoughts in order to transform my reality. (You can read about her journey in her book *Finding Joy After Loss*.) But, still, the struggles continued to pile up.

My business was spiraling downward even as I was

working eighteen hours a day. My father, the person I look up to the most, had been a counselor for my entire life, so I told him about it all. I told him about the punishing hours and how much I missed my family and how hopeless I felt because of everything I had lost. Unfortunately, my father's dementia was progressing to the point where he could barely understand me. This was a huge blow.

The only thing I had left was my own mind.

I took my sister's advice and started reading everything I possibly could about meditation and changing this awful reality I faced. I couldn't just win the lottery to make it all go away—or at least I didn't believe I could at the time—but I could overcome my own mind before losing it to addiction and despair.

I had studied neurolinguistic programming in the past, so while I thought I had meditated before, I learned that all I had ever done was guided self-hypnosis. I went overboard on studying everything I possibly could about what meditation really is. I learned about transcendental meditation, presence, and reality. I learned that the first thing I needed to come to grips with was personal responsibility. I learned that reality was something I created. I learned that it was my own fears and thoughts that had really ended my relationship, and that the despair I was now feeling would only come back to me as if I was a mirror. I had to take responsibility for it all.

Alongside an ongoing practice of radical responsibility—to the point where, if I was in the backseat of an Uber and a car rear-ended us, I would have to take responsibility—I also began to practice this new understanding of meditation.

Over ninety days, I practiced extending the amount of time I could be still and quiet the noise of my monkey brain. It started with just a few minutes of stillness. My body was addicted to the despair and wanted to fight back against the calm. Slowly, gradually, I worked up to thirty minutes of deep stillness.

And then, on day seventy-seven, it seemed like a veil was lifted and the universe itself opened up to me.

With my mind slowed and focused on my breath, my spine aligned, sitting in complete silence, openness, and receptivity, this beautiful moment came through me like a flash of information. Limitless possible futures were open to me. I could see that everything was going to be okay. My kids and I were going to be fine. In this one overwhelming, ecstatic moment, light came into my life.

I had to keep going, to try for more. Every time I meditated, I reached a new level. I forgave the mother of my kids. I found love for her even after she left me for someone else. I forgave my mother. I learned to change the deep beliefs that plagued my subconscious mind. The changes felt incredible, but I still wanted more. I wanted to understand what was happening, to answer the skeptic in me that continued to cling to some degree of doubt for every potential manifestation.

I read *Reality Transurfing* by Vadim Zeland who

discusses a vast informational structure that has the information about multiple possible realities available for us to access. I voraciously devoured *Quantum Jumps: An Extraordinary Science of Happiness and Prosperity* by Cynthia Sue Larson, which outlined the very real science of moving into entirely different realities and witnessing magical shifts in reality.[1] I read Frederick Dodson's groundbreaking book *Parallel Universes of Self*, which outlined a transformational way to use parallel universes to experience any possible reality. I became fascinated by Burt Goldman's *Quantum Jumping*, which described the author consciously quantum jumping into parallel realities through meditation. Intrigued by the connection to physics, I dove into the quantum world, and was met with an astounding array of unusual ideas distinct from metaphysics.

Quantum physics does not rule out the possibility of multiple realities, in fact one of the interpretations, the many worlds interpretation (MWI), literally relies on every possible outcome of every possible decision existing somewhere out there. So, technically, all of this *could* be possible.

My meditation practice shifted to focus on these parallel realities, and the idea that every single infinite version of yourself exists. That if you want to get out of

---

[1] You can check out my interview with Cynthia Sue Larson to learn more: http://www.therealityrevolution.com/interview-with-cynthia-sue-larson-physicist-life-coach-and-writer-ep-98/.

a situation, you have to access the alternate version of yourself that has done so.

I started with my business. My lowest point had been when my kids came to visit and I could not be with them. Instead, I sat in my office for eighteen hours. At that time, I had been focused on the business loan I had to pay off. Once I shifted my focus away from debt and on to prosperity, enough customers and opportunities opened up within just a week's time that I went from working eighteen hours a day to three or four hours a day. Soon, the entire loan was paid off. Not long afterward, I met and fell in love with my beautiful girlfriend. We booked a trip to Hawaii. In every area of my life, incredible things started happening and, most amazing of all, they *kept* happening.

The most powerful moment of all happened during a meditation in which I focused on the pineal gland. Through my research, I learned that the pineal gland is one of the keys at our disposal to perhaps access these realities. The pineal gland has historically been associated with awakenings, and may allow us to see energies and realms beyond our normal senses. At some point during my meditation, it felt like I shrank down into my pineal gland, as if it were a room around me, in which all of these versions of myself existed. The place where I found myself was like an interdimensional gateway where all realities could interact. There was fat Brian, aged Brian, married Brian, childless Brian...all of us were in this shared space where we could communicate with each other. There were timelines so very different that I could never access them and timelines where I could see a path to unimaginable futures. It was as though massive amounts of information were downloaded into both my conscious and subconscious mind.

I came out of the meditation crying and changed forever. I wanted to share it with someone. I wanted to tell people, but who would believe me? It could have just as easily been my imagination and nothing more. Sometimes I still wonder if I created that vision in order to transform myself. I cannot fully know.

What I do know is that the night the Broncos won the Superbowl, I was not fatally-wounded Brian. I was not bleeding-on-the-kitchen-floor Brian. I knew everything would be okay, and it was. I had seen that Brian too, and I knew how to get there.

A COMPLEX CONCEPT OF EXISTENCE

Let's say the many worlds interpretation is real. I like to envision the multiverse like this: imagine there are an infinite number of highways stacked one on top of the other. Each highway comes from and moves toward a specific destination. Branching off of each of these infinite highways are side roads that will move you in different directions. Some of these side roads synch up with other highways that are coming from and heading toward different directions, ultimately taking you to a new destination. Some of these roads cross back over one another.

All of these roads exist simultaneously, and each is just as real as the others. However, you are only observing and aware of the road you are traveling down.

Each highway is available to you, but you ultimately choose where you are heading. You can steer in a particular direction by using your powers of intention, observation, and awareness. Or, you can take your hands off the wheel and let the car go where it will, essentially giving up your control and power.

In the pages that follow, I will show you how to keep your hands firmly on the wheel so that you determine the direction in which you are headed. I will also demonstrate how you have the power to identify those side roads that may offer a better route to your intended destination than the path you are currently traveling. In essence, you will learn to use your internal GPS to lead you down roads that you may believe are inaccessible to you.

Much like highways, I believe all of these different multiverses or parallel realities are accessible paths we can take. We each have the capability of existing in any reality and even switching from one to another. In other words, you can choose the reality you want to exist in because that reality always exists, whether you are observing it or not. It is for this reason that the Law of Attraction—which holds that our thoughts and beliefs control our reality and, thus, what we attract to us—is a powerful and very real form of manifestation. The problem is that most of us do not understand that the reality we desire is already in existence. Until we truly understand this, we cannot achieve the mindset required to get us there.

There are countless and constant philosophical and mathematical discussions about the exact way the universe works. While it is a fascinating topic that we will explore to some extent in this book as well, the truth of the matter is that the details do not matter to us. You and I are here in this reality, right now. What actually matters is understanding that there are endless possibilities for each of us.

You are not stuck. Your future is not determined by chaos. You have a choice, and each choice you make is important. Each decision not only affects the way you move along your highway, but also whether or not you remain on the highway upon which you are currently traveling.

No, I do not have all of the answers about the way in which the universe works. However, I can tell you my experience and that of my clients. I can tell you about maneuvering into a place where credit scores change overnight, snapped ankles were never injured in the first place, alcoholism has no hold, and literally anything is possible.

When you absorb all of the implications of what I'm telling you, anything at all is possible. This is the power of a hypothetical multiverse, and it is at your fingertips.

In the chapters to come, I will explain some of the theories that helped give my experiences credence, and I will share with you some of what I believe to be true. We'll look first at quantum physics and what the many worlds interpretation shows, then what I believe to be our power to navigate these potential, alternate realities. Maybe it will resonate with you, too. Maybe, if only for one of you, it will change you. Maybe it will save your life, like it saved mine.

PART I

REALITY ISN'T WHAT YOU THINK IT IS

CHAPTER ONE

QUANTUM MECHANICS

"This is not the real reality. The real reality is behind the curtain. In truth, we are not here. This is our shadow."

—RUMI

The word *quantum* has been ridiculously overused of late. There's a Quantum Realty in my neighborhood. I've driven past a Quantum Coffee. When we think of quantum jumps, it's often in the sense of a large advancement, such as a "quantum jump" in a business. People are using quantum in everything, and that has diluted some of the word's meaning.

What does quantum actually mean? A term coined by physicist Max Planck in 1900, it is, formally, "a discrete quantity of energy proportional in magnitude to the frequency of radiation it represents." Which basically means that physical properties of subatomic particles are quantized, like a photon is a *quantum* of light. This discretization is something very different from what we observe every day—for example, we don't fill glasses two milliliters at a time, we pour liquid into them continuously. Yet due to the discovery of this quantized nature, doors opened to a whole new understanding of physics, and thus *quantum theory* was born.

In physics, the term quantum jump goes back to Niels Bohr's understanding of electrons and how it has changed over time. He looked at the electrons moving around an atom like the orbit of a planetary system, but we now know that the electrons exist more like a cloud. An electron does not just move; it blinks out and appears in a new location. According to James Schombert, a physicist at the University of Oregon, "Atomic electron transition is a change of an electron from one energy level to another within an atom. It appears discontinuous as the electron 'jumps' from one energy level to another, typically in a few nanoseconds or less. It is also known as an electronic (de-)excitation or atomic transition or quantum jump."[2]

Flippant branding aside, we are living in a quantum age, where we see the fruits of quantum physics everywhere, from our computers to large-scale innovation. You see quantum physics at work every day, without even

---

[2] James Schombert, "Quantum Physics," University of Oregon Department of Physics, http://abyss.uoregon.edu/~js/cosmo/lectures/leco8.html.

realizing it. For example, quantum physics is responsible for how fluorescent lights work.

By using new discoveries in quantum physics, quantum computers will change how we see everything. Quantum computers offer ways to tap into quantum properties—such as superposition and entanglement—as powerful computational resources, by making use of qubits, the quantum analogs to classical bits. They have the strong potential to outperform the classical computers we use today. Theoretically, a computation that may take years on a classical computer could require only a few minutes on a quantum computer. This makes simulations possible that are otherwise too complex to handle within reasonable amounts of time on classical computers, such as complex molecules in the pharmaceutical industry or air flow models in aviation. Due to its ability to perform simulations that are impossible with classical computers, quantum computing will likely change everything from the way the stock market functions to every aspect of information security, weather, forecasting, and trend analysis.[3]

Quantum computers are believed to work in parallel universes by the physicists who believe in the many worlds interpretation of the universe. David Deutsch, a groundbreaking physicist, says that this technology will allow useful tasks to be performed in collaboration among parallel universes. Deutsch writes, "The quan-

---

[3] Cynthia Sue Larson, *Quantum Jumps*. (Cynthia Larson, 2013). 5–6, Kindle Edition.

tum theory of parallel universes is not the problem—it is the solution...It is the explanation—the only one that is tenable—of a remarkable and counterintuitive reality."[4]

As our technologies grow, so does our understanding, which for some physicists broaches the metaphysical world as well, and the mindset inspired by this remarkable potential. In *Quantum Jump: An Extraordinary Science of Happiness and Prosperity*, author Cynthia Sue Larson notes, "The Quantum Age takes us far beyond the Information Age to a way of living in which we delight in uncertainty, thrive in entanglement, and flourish with an awareness of many possibilities. In the Quantum Age, we depend upon the fact that many possible realities coexist at every point in space and time."

> The ideas I'm going to present in this book are of a fantastic nature, but in actuality they have very practical applications and results. People who have practiced these techniques have experienced change in inexplicable ways, literally before their eyes. This is powerful information that will affect the way you maneuver through the world, through whatever is brought to you, and into the reality you are creating.
>
> It absolutely can change your life, so it's important to understand the mechanics of it before you start tinkering around.

---

4   Rivka Galchen, "Dream Machine." *The New Yorker* (2011). https://www.newyorker.com/magazine/2011/05/02/dream-machine.

THEORIES OF THE UNIVERSE

The most readily accepted theory, at least in mathematics, is the Copenhagen interpretation of quantum physics, described by Niels Bohr. He suggested that quantum particles exist as waves, which might be in any place at all until the wave function is collapsed. Each particle is equally distributed in a series of overlapping probability waves in a superposition of state, and an observer is required to ensure that quantum choices are made. Upon measurement, only one outcome takes place. Essentially, the wave function collapses to a single outcome. Whereas in the MWI, all possible outcomes are realized, only in different realities.

As far back as the 1950s, Hugh Everett proposed that every possibility is inherent and that each wave function is real—so that they are all happening. His many worlds interpretation was that possibilities become actualities with each measurement that is made, and that infinite, but slightly different, realities come into existence as each quantum event is observed. In the multiverse, all possibilities are equal and parallel universes coexist, side by side, undetected by one another.

You might have heard the analogy from Michio Kaku, who wrote a book called *Parallel Worlds*, where he uses the example of a stage with trap doors. When you go through a trap door, you drop onto a stage that exists right below the other. All of these stages have a story going at the same time. When you make the decision to drop through that trap door and move in a particular direction, you start to create a newer universe—a new storyline on a new stage—as you move along that path.

To dig a little bit deeper, I like to use the highway analogy that I presented in the introduction. Each highway comes from a previous destination and moves toward another destination. Each highway runs side by side, and each of us is on a particular one. The branches of the highways are constantly growing and moving out indefinitely, sometimes crossing back over each other like exits or side roads. We might sync up with another highway that has moved parallel with us from another location in the past, or we might jump into something completely different as one highway branch mixes in with another.

Now, my highway analogy fits one theory really well—the concept of the multiverse, which is defined by many different theories and interpretations. Each angle of the many worlds view will come at the multiverse a little bit differently, and my goal in this book is not to give you a singular way that the multiverse works. You might believe that we're on all of the different roads at the same time, that we have all of the information about all of the roads at the same time, that we already know the destination for each road, or that we're all moving into a different destination on each road simultaneously. There can be multiple answers, but the key point here as we settle into this concept is that the math behind the many worlds theory has yet to be disproven. In fact, a number of prominent physicists, such as David Deutsch, believe in the many worlds interpretation.

In *The Quantum Problem*, Adrian Kent says, "Every time any of us does a quantum experiment with several possible outcomes, all of those outcomes are enacted in different branches of reality, each of which contains a copy of ourselves, whose memories are identical up to the start of the experiment, but each of whom sees different results. None of these future selves has any special claim to be the real one. They are all equally real, genuine, but distinct successors of the person who started the experiment. The same picture holds true more generally in cosmology. Alongside the reality we currently inhabit, there are many others in which the history of the universe and our planet was ever so slightly different; many more in which humanity exists on Earth, but the course of humanity was significantly different; and many still in which nothing that resembles Earth or its inhabitants can be found."

In David Bohm's holographic interpretation, he proposed the universe as a giant hologram, containing both matter and consciousness. According to this theory, even the smallest fragment of space contains information about the whole.

Meanwhile, the incredible Nobel Prize-winning physicist Richard Feynman advocates what he calls *multiple histories*, which is similar to the many worlds theory. This was an early suggestion by H. Dieter Zeh in 1995. His argument is that a decohering, no-collapse universe avoids the necessity of distinct macro realms. This is through the introduction of a new physiological parallelism, in which individual minds supervene on each noninterfering component in the physical state. This became the most natural interpretation of quantum mechanics—and of course is more than a little bit spacey.

There is also the concept of bubble universes, with a powerful argument in its favor. The idea is that the Big Bang created our universe as well as an opposite universe—moving in the reverse by both direction and time. In that universe, the theory goes, they would look at us as though time were moving backwards. Other universes could have different rules of physics, gravity, time, and everything, with each encapsulated in its own bubble.

QUANTUM PHENOMENA

Throughout this book, I will refer to a number of quantum phenomena, including quantum coherence, entanglement, tunneling, superposition of states, and teleportation. I will also refer to the existence of multiple realities in the form of the multiverse or parallel universes—terms that can be used interchangeably. But before we dive into how this perceived reality may not be the only one, let's establish what all of these terms mean.

QUANTUM ENTANGLEMENT

Apart from his infamous cat, another concept that Schrödinger introduced is what happens when what is done to one particle affects another, even when they are separated by a great distance. Even if they're light-years away, he posited, an entanglement can still be found, though as with all quantum states, it can be a fragile connection.

As the math started to work, experiments began to create these entanglements. Soon, they could entangle two particles and put them in separate rooms. When one particle was moved, the other would move at the same time, as though they were together.

These experiments take us all the way back to Einstein's belief that light speed is the upper limit of speed. So, when Schrödinger observed something ten light-years away moving at the exact same time as another particle, he called it *spooky action at a distance*. It appeared to be moving faster than the speed of light, when in reality there is a synchronization that seems to occur even when entangled particles are separated over a distance of space.

THE OBSERVER EFFECT

Quantum physics is based on an invasive theory, which is that simply observing a system alters it. This has been established in the pivotal double-slit experiment, which involves projecting quantum particles through a barrier that has two small holes in it, then measuring the way they're detected after they come through the opening. A simple yet thorough description comes from *Plus* magazine:

> One of the most famous experiments in physics is the double-slit experiment. It demonstrates, with unparalleled strangeness, that little particles of matter have something of a wave about them, and suggests that the very act of observing a particle has a dramatic effect on its behavior.
>
> To start off, imagine a wall with two slits in it. Imagine throwing tennis balls at the wall. Some will bounce off the wall, but some will travel through the slits. If there's another wall behind the first, the tennis balls that have travelled through the slits will hit it. If you mark all the spots where a ball has hit the second wall, what do you expect to see? That's right. Two strips of marks roughly the same shape as the slits.[5]
>
> Now, what would you expect to see if you used electrons instead of balls? The natural answer is a similar pattern to that of the balls, because electrons are subatomic particles, right?

---

5   Marianne. "Physics in a Minute: The Double Slit Experiment." +*Plus Magazine* (2017). https://plus.maths.org/content/physics-minute-double-slit-experiment-0

You'd expect two rectangular strips on the second wall, as with the tennis balls, but what you actually see is very different: the spots where electrons hit build up to replicate the interference pattern from a wave.

How can this be? One possibility might be that the electrons somehow interfere with each other, so they don't arrive in the same places they would if they were alone. However, the interference pattern remains even when you fire the electrons one by one, so that they have no chance of interfering. Strangely, each individual electron contributes one dot to an overall pattern that looks like the interference pattern of a wave. Could it be that each electron somehow splits, passes through both slits at once, interferes with itself, and then recombines to meet the second screen as a single, localised particle?

To find out, you might place a detector by the slits, to see which slit an electron passes through. And that's the really weird bit. If you do that, then the pattern on the detector screen turns into the particle pattern of two strips, as seen in the first picture above! The interference pattern disappears. Somehow, the very act of looking makes sure that the electrons travel like well-behaved little tennis balls. It's as if they knew they were being spied on and decided not to be caught in the act of performing weird quantum shenanigans.[6]

---
6   Ibid.

This begs the question: is the particle—the light or the paintball or the electron—conscious of the extra barrier? Is it behaving as it is because the experimenter has consciousness of it? The question of consciousness becomes even more prominent when we bring observation into the equation. When in between the barriers, the electron moves into a superposition of state, in multiple places at one time as only a wave can be. But when the detectors were up to observe this position as it passed through the slit, the wave would collapse into a particle again.

In other words, there is an observer effect. The particle can exist in multiple places at the same time, and under observation, it changes. This is the concept of Schrödinger's cat—the unobserved cat in a box that is both alive and dead until it is observed. Scientists realized an electron could exist in a superposition state (unobserved in a box), where it is multiple things while unobserved and a single thing when observed.

Think about playing a video game. The parts of the game world that you're not in don't really exist—you only see the part of the world that you're in. The other parts are in the program, but only the area you're in becomes real. This is the observer effect—it's creating reality as other parts of it collapse away.

If we consider the power of the observer, then we have to acknowledge that every single discussion about the universe involves the observer. Over a long period of time, each observation and belief will reveal some basis in reality. From what I have found in myself, the people I've coached, and the people I've read about and examples I've seen, a more concentrated belief and application of these principles create a larger amount of change—a stronger observable reality—that is massively powerful.

QUANTUM TUNNELING, TELEPORTATION, AND SUPERPOSITION OF STATES

Tunneling suggests that particles have a finite probability of crossing an energy barrier, such as the need to break a bond with a particle when its energy is less than a barrier. Because matter is both particle and waves, it can move through the barrier as a wave then become a particle on the other side—in other words, the wave form allows it to move through barriers and limitations with energy. This explanation of waves is still very rudimentary and beyond the scope of this book to delve into more thoroughly.

Frankly, I still do not understand this concept entirely, but I have been told that "if a quantum particle is trapped within a hypothetical box, it is possible that it has a non-zero probability of being *outside* the box." And so it might move there!

Superposition is also fascinating and difficult to explain in layman's terms. Superposition means placing multiple things together so that they coincide. In the context of quantum physics, a particle can be in a superposition of multiple states. For instance, a qubit can be in a superposition of 0 and 1. Superposition is a powerful tool because if you perform a function on a single qubit in a superposition it's like you're performing the function on both 0 and 1—two for the price of one.

QUANTUM COHERENCE AND MACROLEVEL EXAMPLES

When two waves are in sync in terms of phase and frequency, they are said to be coherent. This does not necessarily imply that all parts of a system are synchronized—there can be incoherent superpositions of states, for instance—but this is the principle of quantum coherence. If you've ever seen a flock of birds all flying together as if they were forming patterns, never bumping into each other, that's a great analogy of quantum coherence.

In March of 2010, Aaron O'Connell and his fellow researchers at UC Santa Barbara witnessed visible evidence of a tiny strip of resonating metal existing in a superimposed state. We're also starting to see viruses with thousands of subatomic particles becoming entangled and existing in superimposed states. Over time, I believe we're going to see more and more examples of this on a macroscopic state.

So far, we haven't seen evidence of many of these phenomena on a macrolevel outside of very controlled experiments. Things that I will explore in this book, such as entanglement between humans, fall within the scope of metaphysics.

THE METAPHYSICS OF TIME AND PERCEPTION

After my home was broken into, I tried to talk to as many physicists as I could. A lot of them wouldn't give me any time at all, but of those who would, many of them worked on a microscopic scale rather than macroscopic. There are differences, to be sure, but research increasingly shows examples of the effects of quantum entanglement, quantum coherence, and superposition of states on a macroscopic scale.

Hameroff and Penrose and the Orch-OR theory of consciousness claim that consciousness derives from deeper level quantum vibrations, inside of the microtubules that are found inside of brain neurons. This may mean that the brain is potentially a quantum computer, interacting with and using quantum signals in the creation of consciousness. This might lend credence to the idea that quantum properties occur on the macrolevel.[7]

Our bodies, cells, and minds are constantly giving quantum signals, which we can visualize as entanglements and quantum coherence. If there is the possibility of electrons and particles being in multiple states at once, the idea is that we can also be in parallel states at the same time—quantum superposition. In the many worlds theory, the wave doesn't collapse. Each of the possible states the particles can take has probabilities associated

---

[7] Elsevier. "Discovery of Quantum Vibrations in 'Microtubules' Corroborates Theory of Consciousness," Phys.org (2014). https://phys.org/news/2014-01-discovery-quantum-vibrations-microtubules-corroborates.html.

with it. If this is true and it can exist on a small scale, then I believe it can also exist on a large scale.

When you become close enough to a person, you begin to share thoughts and feelings. Something might come up and the other person will give you a call right then as it is happening—there is something scientific happening there, which Abraham Laszlo explored in *Entangled Minds*. We also see it in New Age concepts of putting out positive energy and seeing it come back to us. In the same way, if you become entangled with negativity on the news or with your close friends, that negativity starts to rub off on you. That's because we're becoming entangled with those mindsets and people.

The catch seems to be that we're locked into our own awareness. It might be the soul or some sense of consciousness, but it keeps us in the particular reality that we are in. Quantum superposition suggests that who and what we are also exists in multiple other states, even if we are not aware of it. You do not have to believe, as I do, that these other states are actual material realities. They can, in fact, exist as a structure in the information space that has not yet been realized but that still exists as a possibility.

THE CHOICE TO CHANGE

The point is that there are constant philosophical and mathematical discussions about exactly how each of the many worlds might function—and that none of it really matters to us. We are in the reality that we find ourselves in right now, and we want to figure out how to hack that reality, maneuver through different decisions, and achieve things that we didn't believe we could.

I only start with these theories because they come from such incredible minds. We're talking about Einstein and Hawking and Feynman, each of whom spent their entire career coming up with groundbreaking equations and theories about the intricacies of reality. The ideas included in this book are not a kooky, New Age philosophy that I'm making up to sell a promise about a better life. These are concepts that the intellectual giants of our time have developed with credibility and grounding in theoretical models that are mathematically valid.

There is plenty of room for doubt about the exact way that the universe works, and there's just as much room to consider various alternatives. Simply understanding that these possibilities exist opens us up spiritually and experientially.

Understand this: we are not stuck.

Our future is not predetermined by chaos. We have a choice, and the more we understand this, the more important our choices become. If you agree that each destination is available to you, what would stop you from choosing to behave differently to manifest them?

THE CONCEPT OF THE JUMP

Parallel universes certainly make up a significant portion of the collective consciousness at this time. We see endless references to parallel reality everywhere from movies and television shows to popular fiction and video games. It seems like every popular movie and TV show from *Avengers End Game* to *Fringe* to *Rick and Morty* is willing to explore the possibility of a multiverse.

There are stories of alternate realities, or *life tracks* as Vadim Zeland calls them, that are absolutely mindboggling. Frederick Dodson talks about how he twisted his ankle, then laid down and assumed he could jump. When he rolled over and moved his leg again, it was completely healed. Dr. Joe Dispenza has multiple stories of people whose tumors have completely disappeared after a three-day conference. The only way I believe these things could happen is if they are literally moving into another body from another life track.

If we have enough energy, I propose that we can "jump" on the macroscopic scale—at the very least, by creating a mindset jump that manifests a better reality, if not a hypothetical, literal jump. I've seen and felt it. Before the home invasion, I had followed Bert Goldman's set of quantum jumping meditations, as well as Cynthia Sue Larson's guidance in *Quantum Jumps*. Both of these authors are dedicated to the idea that we can build up enough energy through meditation, qigong, and yoga to jump into a parallel timeline. (While I will primarily focus on shifting and maneuvering realities in this book, I have also included more details on quantum jumping and some ways to practice it, for those who are interested.)

It's important to understand that the original road doesn't cease to exist if we take another one—nothing ever ceases to exist. We're observers who can only perceive one reality at a time, or who can only experience one highway at a time, so it seems like we've only ever been on one road. Sometimes, we can be on a road for so long that we forget how to drive anywhere else. Our understanding of driving on a dirt road or a rocky path can literally stop us from being able to switch to any other kind of path. It's more difficult to move in those cases, but it's still possible. The catch is to be able to let go of our mode of transportation and just keep steering, no matter what it feels like.

This concept is so moving, so life-changing for me personally, that I had to write about it—not because it's easy, because it's not. Some of these presuppositions are difficult, and you have to work hard at it and show some grit. But what if it's possible? What if we can move into a reality where our healing and success is real, right now? That's profound. It's powerful.

Whether we're incredibly aware or hopelessly stuck on our own path, the only way to move in a particular direction is with intention and awareness. Most of us just take our hands off the wheel and let the car drive, even if it goes off the road onto another highway. We act as though we have no control at all. The argument I will make in this book—the truth that I firmly believe—is that you can and should put your hand on the wheel. Watch the road. Follow your internal GPS. If you can see a side road that your GPS tells you will be a better route, take it and get to the destination you want. In the pages to come, that's exactly what we're going to explore.

A NEW UNDERSTANDING OF TIME

One of the mistakes that we make when we think about time is to assume that it is linear. We think that we're walking along a single timeline, from points A to B to C to D. We know that time is relative based on Einstein's theory of relativity. The more theories we consider, the clearer we see that time is more like a lake than a straight line. The actions we take are like pebbles tossed into that lake, with the wave going out into the future and the past, which all coexist at the same time as the present.

As Dr. Kristie Miller, the joint director of the Centre for Time at the University of Sydney explains it, the block universe theory says that we can look at our universe as a giant four-dimensional block of spacetime, containing all the things that ever happen. In the block universe, there is no "now" or present. All moments that exist are just relative to each other within the three spatial dimensions and one time dimension. Your sense of the present is just reflecting where in the block universe you are at that instant. The "past" is just a slice of the universe at an earlier location while the "future" is at a later location.[8] Vadim Zeland expands on this idea proclaiming that there is not just one future but multiple futures available to choose from in this block universe.

When I first started to explore this idea of cognition of the future, I found a book called *The Mysteries of the Unexplained*. The author, Adrian Dobbs, was a mathematician and physicist at the University of Cambridge in 1965. He proposed that, as events unfold, there's actually a relatively small amount of possibilities for change at a subatomic level. However, in the process, disturbances are caused in another dimension of time, which then create what he called a psi-tronic wavefront. This can be registered by the brain's neurons, especially for particularly sensitive people or for people who have worked to develop that sensitivity.

Dobbs also used a body of water as a metaphor, picturing a pond and a toy ship launched on one side of it. On the other side of a pond is a very small person who is unable to see the ship when it launches. As the ship travels forward, the waves that it makes begin to reach the shore where the person stands. These waves travel around weeds, leaves, and logs that are either fixed in

---

[8] Paul Ratner, "New Controversial Theory: Past, Present, Future Exist Simultaneously" Bigthink.com (2018). https://bigthink.com/surprising-science/a-controversial-theory-claims-present-past-and-future-exist-at-the-same-time.

place or that slowly drift across the pond as well. Each of these objects creates disturbances in the wavefront.

If that person standing on the shoreline has a lifetime of experience in the things that might disturb a pond's surface, he will be able to note the fine details in the waves that reach him. Through these observations of the wavefronts, he can obtain an image of the objects on the water, and can even calculate how long it will be before they drift to shore.

When people get feelings or hunches or instincts about the future, Dobbs believes that we are actually locked into an awareness of all of these life tracks—the waves—in parallel universes. These are the people who didn't get on the planes on 9/11, who didn't walk into the building that caught on fire, who didn't get on a train that went off the tracks.

In his book *Parallel Universes: The Search for Other Worlds*, Fred Alan Wolfe declares, "The fact that the future may play a role in the present is a new prediction of the mathematical laws of quantum physics. If interpreted literally, the mathematical formulas indicate not only how the future enters our present but also how our minds may be able to sense the presence of parallel universes."[9]

Vadim Zeland's reality transurfing model uses the concept that all time exists at once. All of the future and past can be accessed through what is called the alternatives space. Zeland says that "anything is possible in the

---

9   Fred Alan Wolf, *Parallel Universes: The Search for Other Worlds* (Simon & Schuster, 1988). 23.

alternatives space." He believes there is an information structure that exists all around us that includes all of the possible variations available for the future. All of the different possible points—all of the events—occur at the same time. Everything that is, was, or ever will be exists at once, and all we have is information about it. If I decide to eat a Big Mac for breakfast, lunch, and dinner every single day, somewhere on that path will exist a very large version of me. But I don't have to experience that entire life track. I can access the waves of the future and jump to other places in this ocean of time.

ACCESS TO FLOW STATE

The hypothesis of this book is that we can become increasingly sensitive to these realities and to what we can do to flow through the pond and maneuver our way onto different paths. Part of the process of success is to get into that *flow state*, and I believe that is where we access glimpses of the future.

Interestingly enough, while a flow state usually creates more productivity, that doesn't mean our brains are more turned on. Stephen Kotler's work on flow state indicated the opposite. When you look at the brains of someone doing really well—rappers who are moving quickly, for example—the amygdala and prefrontal cortex has actually shut down. The ego or higher self that makes sure we survive acts like a filter that stops us until we can turn it off and move into flow. In those states, we can access parallel realities of the future that are truly incredible.

The concept of time happening all at once is one explored by many religions. In an Indian text, the idea of Indra's net explains an infinitely large net of chords owned by a powerful god, and he hangs that net over his palace. On each chord is a jewel, and each jewel is a reflection of all the other jewels. Time is like that multifaceted jewel, hanging all around us, that we can gain access to at any time or place.

The concept of prayer could be connected to this idea of time as well. In Lynne McTaggart's book *The Intention Experiment*, she explains how one man wanted to disprove the way studies come to conclusions, so he had people pray for people they didn't know. He gave them a list of people and instructed them to pray for them to get healthier. In the end, he found a significant correlation between the people they prayed for and their health changes, though the prayers happened five years in the past. There's a real chance that the prayers of the past actually did make people better.[10]

Similarly, I believe we can affect our past from the present. Try to reimagine something you did in the past, but this time do it differently. While our beliefs can limit the extent of change that we see at first, at the very least

---

10 Lynne McTaggart, *The Intention Experiment: Using Your Thoughts to Change Your Life and The World* (Atria Books, 2008).

it's a therapeutic exercise that can create alterations. In my own past, I made mistakes and did things that I was ashamed of, and when I went back to alter them, I found I had more confidence. Those things didn't bother me as much. Because we make so many decisions for the future based on things that happen in the past, this is at least a psychological exercise in letting go, if not something real that's actually happening in the past, just across the pond.[11]

The key to effective moving quantum jumps is to let go of your past. As long as you carry with you everything that you are, defined by your past, you'll be limited to what you can have in the future. So I've started to apply these principles in relationships, in interactions with my parents, in childhood traumas...after the home invasion, I wondered whether the voice I heard was my own. Did I go back in time to help myself? I decided that I should try it, just in case. So I got into a deep meditation and went back to that moment. I walked over to myself laying on the couch and told myself to get up and shut the door.

I don't know whether it was me from the future guiding myself, but I do know that I felt like I was being guided. The possibility exists that in that moment, I went back to change the event that happened, and that I'm still alive because of it.

When I go back to change what happened, it lifts me.

---

[11] Try my time travel meditation to experiment with this: http://www.therealityrevolution.com/guided-meditation-the-time-travel-meditation-ep-56/.

It takes away burdens and baggage and gives me the confidence I need to access the things that I've wanted to do.

THE BELIEF FACTOR

The key to all of this lies in what you believe. There is magic in believing.

Going all the way back to birth, we're given beliefs about light and dark, sustenance and satisfaction, happiness and pain. There is a blank slate, and it's filled with what is right, how we communicate, and how we eat—all given to us beyond any choice of our own. They exist outside of us, while our subconscious mind commands our hearts to beat, our lungs to breathe, and our minds to hold the structure of our beliefs. The key is to become aware of the subconscious, apart from what was given to us.

We are formed by these belief patterns. Wayne Dyer says that you need to believe your heart's desires if you want to see them manifest in your life.[12] Proverbs 23:7 says, "As he thinketh in his heart, so is he."[13] Gregg Braden, the author of *The Divine Matrix* and *The Spontaneous Healing of Belief*, says, "To fully awaken our power,

---

[12] Dr. Wayne Dyer, "Manifesting 101: Mastering the Art of Getting What You Want," Drwaynedyer.com, accessed December 2019. https://www.drwaynedyer.com/blog/manifesting-101-mastering-the-art-of-getting-what-you-want/

[13] Proverbs 23:7, King James Version

however, requires a subtle change in the way we think of ourselves in life—a shift in belief."[14]

This is why, before anything else in this book, we had to discuss quantum physics. This doesn't have to be mystical. The scientific rationale can undergird your beliefs to free you to choose the reality you desire.

However, the key is not to understand but to break apart your strict beliefs. To be open to the possibility of new beliefs in your life. Maybe you've noticed reality shifts already, but wrote it off as something else. For example, have you ever had things disappear in your house? Cynthia Sue Larson ran a survey that found 33 percent of people have seen plants or animals disappear. She also talked about her and a friend seeing a sundial on the Berkley campus that looked like it had been there forever, but they had never seen it before.

We're constantly seeing proof all around us. It's just that the beliefs we hold don't often make room for another view of reality. After all, who knows what's really happening here? How would we really know if something had changed or if reality had shifted? There's no way for me to prove any of this to you. All we ever really have is the moment that we're in now.

You don't have to believe me that time is not linear or that we can access other selves. You don't have to believe me at all! I'm a science fiction junkie, not a scientist. I am clearly biased. There might just be a deep part of me who

---

[14] Greg Braden, *The Spontaneous Healing of Belief*. (Hay House, April 2009.)

wants this to be true. But I believe that you can apply this anyway, and it's belief that undergirds all of this.

Belief is what guides us. It's the gasoline in the car on the highway we're driving down. Claude Bristol joins the crowd of experts and philosophers who believe beliefs define us. Our biggest barrier and our biggest tool is our belief.

In the same way sound creates visible waves as it travels through a droplet of water, belief waves ripple through the quantum fabric of our universe to become our bodies—to become healing abundance and peace, disease, or the presence or lack of suffering in our lives. Just as we can tune a sound to change its patterns, we can tune our beliefs to either preserve or destroy all that we cherish.

That belief—that we can choose our own reality—is a heavy one. It forces us to take responsibility for what we have created. Maybe it's this weight that you struggle against. Maybe you're driven by your religious or political beliefs, always looking for things that define our reality and discarding whatever contradicts it.

I can't promise you definitive answers—I can only ask that you don't refuse to believe. Don't be so rigid that you ignore the part of you that is yearning for this to be true. We are never more than a belief away from our greatest love, deepest healing, and most profound miracles.

CHAPTER TWO

BLOCKS TO REALITY CREATION

"All limits are self-imposed."

—ICARUS

Our brains are constantly receiving and transmitting quantum information. Every perception of things like light and heat come from atoms and electrons, which are quantum particles moving in and out of the brain at all times. There has been much research around consciousness and intentions—how we use thoughts, how we transmit them, and how they affect the world around us. The idea here is that the brain is a multidimensional interface that is aware of information on multiple structural dimensions. This brain model was produced by a team of researchers for the Blue Brain Project, a Swiss research initiative devoted to building a supercomputer-powered reconstruction of the human brain. The team used algebraic topology, a branch of mathematics used to describe the properties of objects and spaces regardless of how they change shape.

As they did, they found that groups of neurons make clicks, and the numbers of neurons in a click can indicate the size of a high-dimensional geometric object. The details are difficult to follow if you're not a mathematician, but their excitement about their findings is contagious: to quote them, "We found a world we had never imagined."[15]

Tens of millions of these objects were found in a small section of the brain, and some networks held structures of up to eleven dimensions. Bosonic's string theory contains twenty-six dimensions, super string theory has ten, and a currently popular M-theory also has eleven dimensions. With some unsurprising synchronicity, that's the same number of dimensions that the structure of the brain contains (eleven).

Now, I cannot say exactly what that means. Few of us can, I would guess. It's difficult to wrap our minds around dimensions, string theory, and all of the mathematical complexities of how mass changes, particles are constructed, and vibrational states exist. Even when physicists talk about the fourth dimension of time—a

---

15 Dean, Signe. "Scientists Find Evidence the Human Brain Can Create Structures in up to Eleven Dimensions." Sciencealert.com, 2019. https://www.sciencealert.com/scientists-find-evidence-the-human-brain-can-create-structures-in-up-to-11-dimensions.

well-accepted idea—it's hard to conceptualize. The book *Flatland* imagined an entire universe that existed on one flat dimension. If one of us showed up in that place, they couldn't conceive of us. We would only be an indistinguishable blur. They could see that something bigger and more powerful was there, but that's it.

If our existence were on that flatland, what would it mean to access the person who could walk over and pick up that piece of paper? It would be incredible. That's what we experience within reality hacking. When our senses cannot comprehend what's happening, connecting with the idea that our reality is changing on any level is amazing. In other words, even though we're only ever accessing a tiny part of the structure of our brains and the physical structures that it is capable of comprehending, we know that there is something more. We're seeing the blur. We don't have to stay in flatland forever.

Like the frequently told analogy of the blind man and the elephant, we can only process certain amounts of information, and those sensory limits are vital. We would lose our minds if we could access our full subconscious— when the heart beats, when we blink, when we breathe, what we've done in the past, what we believe, and every bit of quantum information being transmitted via light, energy, emissions, particles, strings, and vibrations.

What we do know is that this information exists, and that we are both influencing and being influenced by it, at all times. The power of the mind stretches well beyond the brain. Before we can look at what that power unlocks, we need to consider what it might block.

THE FORCE OF PENDULUMS

The first block that we are often unaware of is one we also have little to no control of. We see it in bacterial colonies, schools of fish, and herds of animals that move together. Vadim Zeland calls the force *pendulums* and frames it this way:

> On the material level the structure consists of a group of people united by common goals and material objects, such as buildings, furniture, equipment, machinery, technology, and so on. On an energetic level however, a structure appears when a group of people think in the same way, as a result of which, the parameters of their thought energy are identical. Their thought energy finally unites into a single current. When this happens, as if in the middle of an entire ocean of energy, a separate, independent energy-information structure is created which is referred to as an energy pendulum. Eventually this structure begins to live its own life and subjugate to its laws the very people who created it.[16]

A pendulum occurs when more than one person

---
16 Vadim Zeland, *Reality Transurfing: steps 1-5*. (Ves Publishing Group, 2012). Translated by Joanna Dobson.

shares thought energy about a group, idea, or thing. Zeland imagines that the pendulum becomes a separate entity whose only goal is to acquire energy. This is similar to the idea of the egregore discussed in occult literature. The egregore represents a thoughtform or collective group mind, an autonomous psychic entity made up of, influencing, and feeding off the thoughts of a group of people.

From childhood, we're taught to obey the will of others, as if we are in a school of fish. We're taught to perform duties to serve our country, to be with our families, to work for certain companies, to choose political parties. We're taught to align to philosophical ideas and to be part of some state. To varying degrees, we all have a sense of duty—sometimes extending into a sense of guilt—around things that are beyond ourselves.

However, when you read books about the Law of Attraction and the power of the mind over reality, everything seems to be isolated to your specific situation. So we meditate and think about what we want to manifest, we write our goals out and think about them with intentionality, but then walk away to get sucked into a political argument or something that demands all of our energy and attention. The pendulum doesn't care whether you actually love or hate that thing—it just wants your energy.

Think about it like sitting on a swing. If you sit down on it, the swing won't move much unless you or someone else pulls back on that swing or pushes you. Even then, it needs energy to keep going. Similarly, pendulums gain strength by members following rules—giving them energy. You are also giving the pendulum energy by opposing it.

This current of energy is like a frequency—a swing in the middle of an entire ecosystem of different energies and thoughts. As groups of energy begin to move together in the same direction, back and forth, an information structure forms, and it becomes a pendulum. While this isn't magic, there is a concept in old magic books dating back centuries that imagines an entity that literally becomes alive by the thoughts of a group of people. This is, more or less, what I believe is happening. We're feeding the energy of the pendulum when we scoot our energy toward it, and that pulls our energy away from decision-making, maneuvering, and manifestations.

The pendulum becomes a block to reality creation that we try so hard to apply, because it's pulling our energy and focus away. We can visualize the thing that we want all day long, but if we ignore the outside forces that are interacting with us and influencing our decisions and goals—Law of Attraction all on its own, for example—then the energy in the pendulum dissipates. The swing slows down.

The pendulum has its own agenda and goals, and it takes your energy by pulling you into a reality that resonates with it. It will pull you into realities that serve its own goals and purposes and have nothing to do with you.

When we're trying to manifest our dreams by maneuvering through these realities, understand that we can never avoid these pendulums. They're always going to be there. We could sit in a dark room all alone without any noise and barely any food, perhaps, but that's not a good life. Instead, we need to be aware of pendulums—the outside forces moving together in a specific way—so that we can overcome them.

PENDULUMS IN EVERYDAY LIFE

Pendulums don't have to appear as entirely negative. I'd love to see the Reality Revolution become its own pendulum, with the intention that everyone can choose the reality that they want in their lives. If we attach to that kind of pendulum, it can guide us and even force us to move in that direction. Every day, we can align with positive or negative pendulums that move in the direction of our future goals or very far away from them.

Online behavior has become such a clear example of pendulums in recent years. We say and think that online interactions are irrelevant or that our discussions are nothing more than opinions, yet we pour our energy into them, and someone else responds in kind. The "internet troll" doesn't care what your belief is—they just want to say something the opposite to what you said to get the energy of your resentment and anger. They find joy in creating resistance by saying something so ridiculous you can't help but respond. When the back and forth begins, they are immersed in their pendulum and sucking everyone else in with them, mindlessly pushing and opposing energy.

We see pendulums in politics and sporting events. We see them in movements related to gender and material wealth. Some of them are great, some of them are disastrous—but there's no way we can avoid them. The important thing is to become aware of the pendulums and how we choose to interact with them.

The most powerful emotion that can feed into a pendulum is the same one that shuts down our awareness: fear. It's a lot easier to be fearful of something than to create a positive emotion around it. Fear triggers the amygdala—the inner brain. It calls all the way back to when we started evolving and had to be acutely aware of fear in order to survive. Even though we're more sophisticated now on a larger level, we're still those same beings that used fear to survive and evolve to this point in the first place. If we don't have awareness, fear often takes over, and that's when we get sucked into the nearest, strongest pendulum.

Politicians use that fear to create energy to force a movement—to push the swing. It's the easiest way to get the energy that they need for forward motion, even if a strong group of people opposes them. The pendulum strives to attract as many followers as possible, making a distinction between its group and all other groups. It wants to be special, to be its own little school of fish. If you outright oppose it, you're giving it energy.

The pendulum will also hide behind masks to make itself look lofty and ideal. It will make you feel honorable for opposing it. For many of us, our first response is to speak up against the things we absolutely disagree with, or that make us angry, or that are terrible or wrong. That's an understandable instinct. But we have to know what it is that we're feeding when we engage. The pendulum will play on our feelings and make us feel guilty or enraged or obligated to duty. It wants that energy. It wants to pull us into wars and imprisonment and loss. It moves us into timelines where our fears actually occur and where we feel justified in engaging. The pendulum feeds our need to feel connected, deep down.

Remember, this is a new way to understand reality. We cannot ignore the pendulums or make them disappear. The battles will always exist. Instead, we need to be aware of the puppet-master effect that these pendulums create as we mindlessly move together in the push and pull of emotion-heavy energy. We need to be aware of the entity that we are creating with our thoughts and energy so that we can move toward positive pendulums that bring about better timelines and futures—not only for us as individuals, but for us all.

THE LAW OF BALANCE

Pendulums are powerful forces that we can see in the natural world and human nature alike, we can also see balance as well.

One concept outlined in *Reality Transurfing* is the law of balance. Everything in nature strives toward balance. If the air temperature changes, it's balanced out by the wind. If you drop something from the top of a building, the object has potential for energy until it hits the ground and becomes a lower form of energy. Wherever there's excess potential of an energy form, nature always wants to lower it.

The law of balance is always trying to move us toward equilibrium.

I have noticed in my life, when I attribute importance, a counterbalancing force will always come into play. Think about it: have you ever noticed that when you want something badly, you can never seem to get it? That is because your creation of importance brings balancing forces into play that work directly against your intentions.

I think of it as inner and outer importance, which represents an estimation (or over-estimation) of your own virtues, powers, shortcomings, or problems. If you decide that you're an important person who does important work, inner importance goes off the scale and the universe will try to show you just how wrong you are. And the opposite is true, as well.

Outer importance works similarly. When you attribute huge meaning to an object or an event taking place outside of you, excess potential is created and things go wrong. Unfortunately, it's much easier to control inner importance than outer importance. This includes all unbalanced feelings. Indignance, dissatisfaction, irrationality, restlessness, anxiety, despondency, embarrassment, fear, or guilt. Attachments, codependence, deep admiration, exaggerated affection, idealization, worship, pride—all of these are forms of importance that create excess potential.

Importance feeds fear, which builds pendulums and is another form of block. You can diminish fear, or you can feed into it and create excess potential and play into the pendulums even more.

The helicopter soccer parent makes their son so important that they go to every event and spend every night and day working with him to become better at soccer—until the son decides he doesn't want to play anymore. A balancing force occurs that ends up taking away the thing that they wanted, and that maybe the son had wanted as well. Sometimes the importance we express with our kids is motivated by fear, which creates balancing forces that put everyone at a disadvantage.

So how do we reduce importance without becoming negligent? First be aware of when a situation has arisen as a consequence of projected excessive importance. This is an intense force, and until you can believe that every single problem is created by excessive importance, that intensity will bury you. Catch yourself. Stop blowing things out of proportion. Be spontaneous, improvise, and relax. Additionally, use humor to minimize importance. Laugh at the situation and the importance you project upon it will naturally diminish.

In fact, studies have been demonstrated that lucky people are really just relaxed people. By minimizing importance, they are actually able to create better luck and have better results.

One study comes to mind that statistically identified lucky and unlucky people, then set up several factors to determine whether it was true luck. In his book *The Luck Factor*, Richard Wiseman details a simple experiment in which he gave people a newspaper and asked how many total photographs were in it. Some participants took several minutes to answer and others went back and rechecked multiple times. All of them could have revealed the correct answer in seconds, because the second page of the newspaper contained a half-page message that read, "STOP COUNTING—THERE ARE 43 PHOTOGRAPHS IN THIS NEWSPAPER." But nobody saw this message because they were so focused on looking for photographs. The newspaper also had an ad that read, "STOP COUNTING, TELL THE EXPERIMENTER YOU HAVE SEEN THIS AND WIN $100." Nobody saw that message either.

Those of us who are relaxed and willing to look at the big picture will see the guideposts that are blatantly placed in our path.

Another test was to bring people to a bar for an interview, but with a $100 bill laying outside on the sidewalk. The lucky people picked them up, while the unlucky people walked right by. Likely, they were so focused on their fear or looking ahead that they weren't aware of their environment. The lucky people, on the other hand, had a sense of lightness and ease about them, in almost every instance.

Instead of luck, maybe it's simply a mindset and habits. The lucky people had ways to relax themselves. They had a pleasantness and an open mind toward different possibilities. They had a lower sense of importance and little need for balance to take them down a notch.

Several religions convey the message to let things go and to detach from outcomes. Perhaps they became aware that importance and focus on results pulled them away from the results ever occurring.

> Whenever you're struggling with something, ask yourself, "Am I attributing a higher level of importance to this than it needs?"

SELF-FULFILLING PREDICTIVE IMAGES

In our childhood, we create images and ideas of ourselves that limit our possibilities. For example, Bob Dylan could have started to believe he had a terrible voice and couldn't sing. In that scenario, we wouldn't be blessed with all of his amazing music. It's not exactly wrong—he doesn't have a great voice. But if that became his belief about himself, the world might be a very different place.

Once you start creating filters in your mind, even if there is truth to some of them, those beliefs can become limiters. We start to look at ourselves from the outside in—through the filters—and begin to see examples around us that confirm it. Soon, the beliefs minimize our ability to do things that otherwise might have been amazing.

Beliefs are powerful. If my son says to himself, "I can't do math," he'll always struggle with math. But if I tell him that I believe he can do math and then someone can show him how, he will get better. An incredible plastic surgeon, Maxwell Maltz, wrote a book called *Psycho Cybernetics* about this very thing. He noticed that he could change just one tiny little thing about his clients that no one else would notice, but it would transform their view of themselves entirely. They would see themselves in a different light and a greater sense of awareness.

The images and ideas that we give ourselves can absolutely block or enable the creation of new realities. When you believe in yourself, you can steer toward a productive and useful goal. When your image of yourself is positive, it's limitless.

When I was younger, I had a huge underbite. I could barely smile without being aware of that underbite, so I kept my mouth closed for pictures. It was a terrible feeling. Finally, it came down to surgery to fix it. They actually broke my jawbone, pulled bone from my hip, and then rewired it all back together. I thought it was worth it.

What I realized, however, was that most people didn't even notice. No one commented on my smile or the change. This thing that I thought was a huge deal externally turned out to be nothing at all. But because my view of myself changed, I was able to be a debater and a public speaker. My internal belief was free of that image and limitation. Often, we don't need the surgery at all. We just need to identify the images that we're creating.

We cannot manifest what we don't believe, and I acknowledge that this kind of power over our existence is difficult to believe. Consider this: if the holographic theory of the universe is correct—and there is substantial theoretical physics and astrophysics evidence that it is—every tiny part of it is a reflection of the whole.

Assuming this is true, then like the gems in Indra's net from chapter 1, every little piece of us contains information about the entire universe. Every single person can affect the universe itself. Every action, behavior, or minor thought that you have or do—when you make your bed in the morning, when you pick up the dishes—causes a ripple that affects the whole. If you can begin to believe that your seemingly insignificant choices have a very real impact on the whole, then some of the blocks to manifestation will begin to clear away.

HOLOGRAPHIC HABITS

Neuroscience tells us that repeated behaviors fire specific receptors over and over again until they wire together to make them easier. Usually, these behaviors—habits—are linked to emotions in our body. This is how we establish our reality. Everyday habits, even if we think they're irrelevant, become a bigger factor than we realize. We have habits around our health, how we talk to ourselves, our relationship with money, our relationships to other people—habits constantly influence our decisions and the way we function on a daily basis.

Most of us want to be healthy, organized, and wealthy, but our habits don't reflect those desires. Like our DNA mirroring the universe, our small-scale habits define us on a larger scale. They are holographic. If our habits are to smoke cigarettes every day and tell ourselves that we're not good enough, then we don't actually care about our health. If we keep a messy room, then we're not important enough to be organized and clean. You might not believe that to be true, but what happens if you see a penny on the street? If you don't bend to pick it up, that's because it's not worth it. You're not open to receiving it because of what you believe about it.

Now, what happens if someone offers to buy your meal? Most of us have the habit of saying, "Oh, no, it's okay. I'll take care of it." There's a part of you that isn't open to receiving.

Some of these mindsets and beliefs stem from generational things that we've learned. My mom would always say, "Sorry, son. Money doesn't grow on trees." Today, I have to consciously remind myself that money is an unlimited resource. Otherwise, when I try to manifest money, I believe there's only so much that I can get, and that habitual belief becomes a block.

I believe we move into timelines that match the reality of who we are. If I say to myself, "I am wealthy" then act with the habits that wealthy people have, I will move into that timeline.

If I look at a hundred people who are in the kind of timeline that I want, and they all have certain habits, then I need to access those habits too. Brendan Burchard's fascinating study that drove the book *High Performance Habits* identified six common habits of the highest performers: clarity, energy, necessity, productivity, influence, and courage. These aren't just about brushing your teeth or picking out clothes, but are actual models that have moved people into timelines of success, wealth, great relationships, and health.

If we see what they are doing and still choose habits that interfere with sleep, don't nourish our bodies, and aren't productive, then we're saying that we can't have those things. We're discrediting the belief that we are successful—that we deserve success—on a deep level.

FEAR IS AT THE ROOT

Recurring within each of these blocks, we see fear showing up over and over again. Fear is the easiest emotion to connect to. It's an automatic response, where emotions like gratitude and positivity take work to create.

Like the pendulums, we can't eliminate fear. It is a warning system that we've been given that has helped us survive. In research settings, when fear is removed people get themselves in dangerous situations. The fear itself is not a problem—it's our lack of understanding around how fear works. Until we understand fear better, it can easily take over and limit our ability to take action.

Neville Goddard, one of the great writers of our time, said that "feeling is the secret" to manifesting. Hacking the reality that you want is connected with hacking your emotions. When you come to a better understanding of your response to fear and your ability to generate the feelings of the life that you want, you open up the potential for a huge shift toward a positive lifeline. Instead of fear, what would success feel like? What does that perfect relationship feel like? How does it feel to be healthy? This is not an easy exercise. Fear will always exist in some way, but you have to work to create other feelings as well.[17]

If a significant event happened in your life, go back to it and analyze it. Talk about it. Look at what ultimately came from it. Then go all the way back to the moment when you felt fear, and adjust your response to

---

[17] I recommend reading the book *Feel the Fear and Do It Anyway* by Susan Jeffers.

it. Incredible things can happen when we reframe past and current circumstances.

The more fine-tuned this becomes for me, the more I can subvert that initial fear response with something else. Once, I was sent an email with a heavy warning about my online business. I was going to lose everything. My business was going to shut down, I was going to go bankrupt, and all of the money I had was going to be gone. Immediately, I said, "What would be my response if I felt okay in this situation?"

I stopped, took a moment, and released that fear.

It's difficult to do, but I made it my intention to avoid the fear and act as if everything was going to be okay. Within ten minutes, I received a follow up email that said, "I'm sorry, we sent that to the wrong person. Everything is fine."

If you want to create a reality, it won't magically happen just because you want it. You have to take action. Action is like a midwife guiding you between where you currently are and what you want to occur. And you can't take action if you are overwhelmed by fear.

We know this. Motivational speakers and self-help speakers have written about fear and talked about it over and over again. We're all aware of the truth. Unfortunately, many of us are addicted to fear just as much as others are addicted to alcohol or substances. Our body's chemicals release into our brains and bodies when a situation comes up, and we experience those chemicals so much that it becomes an addiction. It's habitual.

Sometimes, when we find ourselves making the same decision over and over in spite of what we want to manifest, an addiction to fear might be the answer. It's our responsibility to break that addiction, change the habits, and start relying on more positive things.

RELEASING LIMITS AND ACCESSING POTENTIAL

Brendan Burchard talks about going skydiving, and how scared he was about jumping out of the plane. In all of that fear, he didn't really hear what the guide said about looking at his watch or knowing when and where to pull the cord. When we jumped off of that ledge, he became so overwhelmed that he didn't remember anything. This happens to us all the time. Fear blocks our ability to make the decisions that we want and need to make.

Fear could have paralyzed me on the night of the home invasion. If it had, there's no way I could have made the decision to shut the door and run away. Of course I felt fear, but I couldn't let it limit me.

Our moments of greatest fear are also our greatest opportunity.

When an Olympic high diver stands on the platform, he can't eliminate fear. But if he lets it take over, he risks taking the wrong step or making the wrong movement. He can't win, even though he has done that routine so many times. If he lets fear become energy, and chooses in that moment to take the actions he knows he can take, the gold medal is just within reach.

What we choose, moment to moment, defines us and our future. Whether it's in relationships or business or anything else, don't let fear make the decisions for you. Instead, turn your awareness of fear into energy. If you can become grateful for the fear because of the energy that it gives you and focus that energy and allow it to move through you, then instead of a block it becomes incredibly powerful.

ELIMINATING OR MINIMIZING OUR BLOCKS

Neurolinguistic programming (NLP) was created by Richard Bandler and John Grinder after watching a famous psychotherapist and hypnotist named Milton Erickson, who also happened to be paralyzed from the head down and could only work with his voice. When people came to see him, they would consistently have these powerful, overwhelming transformations in their lives. When they broke down the patterns of speech he used and the way he accessed their emotional states and began to model it, they arrived at the concept of NLP, and it continues to change lives. Tony Robbins, for example, is who he is today because of this system.

NLP was actually the topic of my master's thesis, and it continues to inform me to this day. The idea behind neurolinguistic programming is that some people are kinesthetic—meaning they filter information through their feelings. Some filter through their senses—visual or auditory. The big discovery in that is that you can access the way people understand information by the way they say their words (auditory), move or hold their bodies (kinesthetic), or move their eyes (visual).

One of the applications of this theory is to identify when someone is lying based on where their eyes go—up and to the side, connected to creative aspects of the brain, imagining what their response should be.

Within NLP there is a concept called anchoring, which works with the way we process and use information to make emotions and concepts "stick" in the brain. This is something people use in storytelling all the time. A person who can tell a great story can get you to access deep emotions. If they want to anchor that feeling, they might touch you on the shoulder in a certain way when they say a certain word—squeeze your shoulder and say *yellow*, for example. After a few times, when they squeeze your shoulder and say *yellow*, it will bring out the feelings from the story, even without retelling the whole thing. We become anchored to that moment and its feelings.

Bill Clinton and Barack Obama use that classic thumb point as an anchor. Persuasive speakers can set that anchor early, then come back to hit you with it later. Advertising uses logos and trademarks. They'll put together an incredible sequence that evokes specific emotions, then show their trademark. The next time you see the trademark, they hope you'll be anchored to the feelings from that ad.

I believe we can spend time in meditation creating multiple anchors to maximize feelings and states, completely on our own. I have an entire routine connected to the directions my eyes look, the connection to my senses, and the memories I experience. I look up and to the left, down and to the right, down to the left, up to the right, and back down to the middle. I think about what I'm smelling and hearing. I connect everything to those eye movements and senses, like the eyes might be a joystick to our souls. In this way, anchors become a way to speed up the process of manifestation and remove the blocks.

ANCHORING IN PRACTICE

The best way to move into a particular reality is to recreate a familiar emotion that you can associate with your goal. Looking at the Law of Attraction model, we continuously find that the most powerful technique to bring about the realities that we want is to anchor them in feeling. When the movie *The Secret* came out, they talked about what it might feel like to be in that experience, but they never explained how to do it. How do you bring up an emotion without it being reactive? And if we want to manifest something we haven't experienced, how do we create emotions for it?

It's difficult to think about what it would feel like to

have that dream home or accomplish that goal. What's interesting is that the feeling doesn't have to be specific. You can reflect a past emotion and project it outward throughout your day, and things will come back to bring those emotions to you.

To create this experience with a client, I would look for an incredible moment in their life—something that brings out a lot of emotions. It could be a song or a situation, a memory that someone has. You can tell when you've found it because their body language shows excitement and joy. When you're doing this for yourself, we can find that joy that can be an anchor.

To empower the anchor, we would then do something extra alongside the song or identifiable emotion. It can be a physical touch, squeezing your finger, scratching your wrist...The first time you do the anchoring movement, it's just barely beginning to create a neural pathway in your mind. The next time that you do the exact same thing while recreating the emotional state, the pathway strengthens more and more.

I like to remember moments with my father, who passed away. I can see us sitting in the stands at the Broncos game when Tim Tebow threw a touchdown pass in the playoffs. We jumped up and down as the stadium roared, and we embraced each other, screaming with joy. I can feel the goosebumps and all of that energy, and when I do, I scratch my wrist a little bit.

The next time I have another powerful, positive

moment, I try to anchor in the same way. Maybe I'll hear a beautiful song that reminds me of a concert with my friends on a perfect day. I'll scratch my wrist a little. Then I might add a particular word that's special to me. Over and over, the impact of that pathway becomes exponential. Eventually, I start to control my ability to create those emotional states.

Again, *the secret* to creating your life is to feel it. Your heart is much smarter than your mind. As you broadcast those feelings out, you tune into their frequency and vibrations, and your life begins to pull toward similar vibrations like a magnet. Many of us watched *The Secret* over and over. We created vision boards and books and set intentions around what we wanted...and then nothing happened. The feeling is the key. Projecting feelings of fear and anxiety, thoughts of scarcity and lack of control, sets your heart and mind off in the wrong direction. Anchoring is a way to tune the body with the heart and begin to open the doors for manifestation.

ACCESSING BRAINWAVES

This is not likely to be your first book on this topic. By the time you make it to Brian Scott—someone you might hardly know of—you've already read multiple other authors. And every one of them will have a section on brainwaves, just like every author has to talk about the double-slit experiment.

Even so, we do need to talk about brainwaves. They are at the root of all of our thoughts and behaviors—the synchronized, electrical pulses coming from masses of neurons in our brains all communicating with each other.

Simplistically, we can think about brainwaves like musical notes. Low-frequency waves are like a deeply penetrating drumbeat, while higher frequency waves are more like a subtle, high-pitched flute. And each of these waves change according to what we're doing and feeling. Slower dominant brainwaves leave us feeling tired, sluggish, dreamy, or relaxed. Higher frequencies make us feel wired, hyper, and alert.

While there are as many as fifty different kinds of brainwaves, there are a few that stand out. The lowest brainwaves are infralow, at less than .5 Hz. These are called *slow cortical potentials*, which underlie our higher brain functions and have evaded most studies. Very little is known about them.

In delta waves, at .5–3 Hz, brainwaves are still slow, and they are actually loud. These generate the deepest meditations and most dreamless sleep. Healers are very much in their best state when they are awake in a delta wave state. This is also where healing, regeneration, and restoration are stimulated.

Theta brain waves, measured at 4–7 Hz, are the brain frequencies of the barely conscious states just before sleeping and just after awakening. Theta waves occur most often at the beginning levels of sleep and are dominant in deep meditations. It's in these waves that I believe a gap is created that gives us access to the universe for milliseconds at a time. That unlimited cosmic memory of information that we're trying to align ourselves with opens up and slips into our brains when we make it into this brainwave state.

At 8–12 Hz, we are in alpha waves, which are dominant when we're relaxed and awake. More than that level becomes beta waves, when attention is directed toward cognitive tasks and fast activities. This is when we're alert, attentive, problem-solving, using judgment, and making decisions. Dr. Church explains that alpha waves are the bridge that connects these brain wave states. "Alpha (8 to 12 Hz) is an optimal state of relaxed alertness. Alpha connects the higher frequencies—the thinking mind of beta and the associative mind of gamma—with the two lowest frequency brain waves, which are theta (4 to 8 Hz) and delta (0 to 4 Hz)."[18]

There's a fascinating level beyond that, though, in the 38–42 Hz range, which Dr. Dispenza theorizes indicates manifestation. These are gamma waves, which sometimes look like they come directly from delta. They are the fastest brainwaves, and they look completely different from all the rest. Gamma waves seem to occur in the states of universal love, altruism, and higher virtues—nothing like high levels of concentration. Dr.

---

18 Dawson Church, *Mind to Matter: The Astonishing Science of How Your Brain Creates Material Reality* (Hay House, Inc., 2018). 72.

Dispenza's work has attempted to identify how we can get into a gamma state with spiritual moments and intentional focus. In his book *Mind to Matter*, Dawson Church states, "Gamma is the highest brain wave frequency (40 to 100 Hz). It's most prevalent at times when the brain is learning, making associations between phenomena and integrating information from many different parts of the brain."[19]

Without much data about people existing in gamma wave states for long periods of time, we do see those states associated with qualities of transcendence, spirituality, and an awakening to creative insight and mental lucidity. There is research showing that the meditative singing bowls, gongs, and chants create gamma waves quickly. The resonating sound can pull you into that state, and of course our examples of high transcendence often come from those traditions.

The idea is that when we're in these states, we can access source information or the space of variations. Even though we don't know it, we're accessing information about possible futures, realities, and ourselves. I believe this is what's happening when we're dreaming because we're in a delta state, and that reaching that state while we're awake and meditating can open that door with more intention. Zeland describes dreaming as an exploration of possible alternative futures and pasts that exist in the space of variations.

---

[19] Ibid.

What you place your attention on can open up a corresponding sector of parallel realities related to where your attention has been placed.

On a small scale, we might focus on something wrong, hang on to our blocks, and open ourselves up to a lifeline full of anxiety and stress. The better you get at controlling your attention and focus, the more you will be able to avoid negative realities. With intentional focus, you can come into possible future scripts that are wonderful and delightful.

The quantum processor of our subconscious limits the full range of information that we can consciously process. Your heart can communicate this information to you through feelings. The next time you have a feeling about doing something, it's likely based on the tons of information that it received while you were in those states of mind. Feeling is the secret.

TECHNOLOGY: DEVICE OR DISTRACTION?

We are in a unique time in history. Everything is happening faster and faster. Consider Moore's Law, which tells us that right now, we are moving exponentially toward incredible futures. We're holding phones in our pockets that contain computers that used to be the size of buildings. We can process massive amounts of information. I hypothesize that we have become in tune with this technology, and it is bringing us into that future. As we place our attention on possible realities, technology is accelerating this process.

Technology isn't something to fear—it has unlocked incredible insights. *The Flow Genome Project* looked at states of bliss and flow and the way our brains can work in certain cases. We're able to follow brain waves and look at the structure of our minds and the universe. None of that information would be known to us if we didn't have the technology that we do.

In Silicon Valley, there is a thing called 40 Years of Zen, where executives pay $15,000 to go to a beautiful location and hook themselves up to an EEG feedback machine. There, they work through forgiveness, emotions, and issues while measuring their brainwaves. When they reach that perfect brainwave state, it's like a state of Zen that someone studying Buddhism for forty years can reach—hence the name 40 Years of Zen.

While a Zen retreat is not feasible for most of us, I do recommend devices that can access where your brainwaves are at. I use one called Muse, which is a brain-sensing headband. When you're in alpha or theta, birds will tweet. When you move into beta, the sounds of thunderstorms roll in. It's fairly cheap at a couple hundred bucks and gives you an active way to monitor your brainwave states. There's also a brainwave starter kit by NeuroSky that can be helpful. A lot of people use Solfeggio tones, which are pulsating brain synching rhythms that train your brain into certain states. Right now, dedicated meditation practitioners are developing guided virtual reality meditations, which could open up a significant increase in our ability to enter into these realities because of this technology's ability to create three-dimensional realities.

Infinite possibility literally means *infinite* possibility.

Now, technology does become dangerous when it becomes a distraction. If it is a block rather than a tool, we should reconsider our relationship to it. In any case, we have choices about how we use it—how we embrace it and align it with the way we're becoming awake inside of ourselves.

We have all been dreaming, as though we're actors in a movie. The script has gone on and on, and we have played our roles. Using this new model of reality, there are an infinite number of scripts available to us. We simply have to choose the script that has our optimal end goal in mind. It's time to wake up from the dream, to realize we're in the story, and to choose the script for the life we intend to live. It's time to decide whether we want to keep going or to change the script wherever we can gain control.

CHAPTER THREE

PLAYING WITH REALITY

At every point in time, there are infinite possibilities and a parallel reality exists for each possibility, so there are literally infinite branches...

—DANIEL JACKSON, *STARGATE*

There was a point in time, before the home invasion, where I truly felt like I had lost everything. My girlfriend left me for the love of her life. The mother of my kids left me for Oregon and took the kids with her. Now, I had nothing left at all. I was devastated.

With nowhere else to go, I moved to California—to the home where the invasion happened—and began to feed addictions to work and to alcohol. I felt like I had no hope, no reason for living, and suicidal ideation began. I remember sitting down with my dad, whom I was quickly losing to dementia, and telling him how I felt.

"Dad," I said, "I've lost hope." Even though he could barely understand me, I would talk to him about my business and my sadness. I would tell him, "There's just... nothing I can do. It is hopeless." The pain was so much I wanted to end it. I was close to giving up.

As I felt myself slipping further down the rabbit holes of addiction and despair, I knew I needed to grab onto something to overcome my mind. I was fortunate enough to find a mentor to help me, and the first thing we did was to establish a morning routine and a practice of meditation.

What I thought I knew about meditation from my neurolinguistic programming study had only been guided hypnosis for myself. Now, I wanted to change my reality. Everything about my chaos was my fault. So my first goal was to calm the chaos. At first, I could only handle a couple of minutes of calming my monkey brain down. Then I just wanted to sit still for thirty minutes without moving.

Every second was a battle. My body was addicted to emotions of despair, and I had to fight my ego and my senses every time they told me to move. But every time I meditated, I came to a new level. I came to grips with responsibility and forgiveness, and with each realization came more struggles with my subconscious mind.

My next step was to silence my thoughts, which felt

impossible at first, but I developed techniques to subvert my thinking brain. I began to meditate on the void and experienced brief flashes of something else behind the curtain. Realms of profound beauty.

With my monkey brain in some form of compliance, my next struggle was against the subconscious blocks that had accrued over decades of time. I wasn't going to be able to unlock those things quickly—it's a difficult process. So, as with everything else in my life, I began to research it. I connected with the New Age community. I slept with guided affirmations and programs meant to release the subconscious.

Finally, I worked to release my natural inclination to doubt. Previously, I would have thought something along the lines of, *All this sounds nice, but it holds no basis in reality. There's no way we can manifest alternate timelines. This is all coincidence.*

When I began to research quantum physics I started to understand the observer effect and the ways that light itself can be both particle and wave. I learned how we are creating reality on subatomic and molecular levels, and most likely on much larger levels than that. I discovered superpositional states and the way that parallel realities exist around us and within us that can be accessed through information fields. How a source field can contain the information of all of existence, everything imaginable, every single action that I could possibly take.

It started off incredibly heady, but the more I learned

the more belief I came to. Soon, I began to do parallel reality meditations. My realities began to clearly change, almost as if I had jumped into an entirely different reality. I paid my loan off in full. I dropped my work hours. I accessed better opportunities. I built relationships with my kids and fell in love and enjoyed this life that I had been drowning in just months before. All the while, I was playing with deep meditative states, for hours at a time, trying to move through parallel realities.

When the home invasion happened, I believe I made a bigger jump than any shift I had experienced in those states. I had embraced the ideas of parallel reality surfing and devoted time to the meditations, but part of me was just having fun with it. I was exploring—maybe it works or maybe it doesn't. I was experimenting to find out. When my life so clearly changed trajectory, the doubt began to melt away. I dove into as much research as I could, this time in complete earnest.

When I started to practice maneuvering through realities more intentionally and wholeheartedly, I found that I could adjust small things—my credit score, for example—or very big things. After struggling to quit drinking, I decided to move into a reality where I succeeded. The next day, I had no urges or desire to drink. There were no withdrawals or psychological effects of stepping back, because I was already in a reality where I had not been drinking. It was never an issue again.

As we begin to talk about parallel reality surfing and,

in the next part of the book, hacking your own reality, understand that I could simply be insane. There is always the possibility that I'm using my own creativity here. For me, that doesn't matter. I'm not talking about getting to the truth. I'm talking about what has truly worked for me.

I took that mentality and began helping other people. I used the basic knowledge that everyone uses—demonstrated throughout my technique and in other surfing and jump techniques in the resources section—and helped them use those concepts in their own lives.

Since then, I've seen people get new jobs, new relationships, overcome drug problems, overcome health issues, wake up without chronic pain...This knowledge and information began to spread, until it made it to this moment, in this book, with you as my reader, on the verge of choosing your own reality.

A COLLECTIVE SHIFT

We are not maneuvering through realities by ourselves. Like the pendulum theory, people are creating realities in groups. Some event in the past might have pulled an entire group of people into an alternate reality, because we gain what we focus on. This is potentially where we see the Mandela Effect being discussed on a larger scale—people realizing that there are differences to the reality they remember.

Because nature always wants to bring balance, it

might be common for an entire universe to change than to simply alter a single person. Group changes are the easiest reality shifts, especially at points in history like this one, where fear is so pervasive, where we are bombarded with Wi-Fi signals and radio waves and TV signals all coming at our brain.

Even then, the pineal gland has an electromagnetic effect, blocking out all of that information in what people refer to as the third eye. It sounds woo-woo, but this little gland in your forehead has a piezoelectric effect. It has water inside of it and crystals, and it has rods and cones like an eye. In spite of everything going around us, it can still give us access to the source.

When we gain access to the source, we gain information about what to be fearful about.

I believe this is why we're at a crux of so much fear. We're ascending, and it's a scary thing. The evidence is all around us. People are more aware. You can start a group on Facebook that says you've unlocked the Matrix and thousands of people will join you to talk about a new awakening. Something is happening, and technology is moving it along at an exponential rate. As technology expands, so do our minds. The pineal gland keeps adapting and changing, responding to the way it needs to protect us and opening us up to new potential.

In other words, our minds are literally changing.

We only have our senses, which only take in fractions of a percent of what's really going on around us. Our experience is subjective, and as we open up to more fear, that voice of despair becomes attractive. We can sink into it, or we can embrace the change and become powerful in it. We can become alchemists to our own subjectivity. We can create a pendulum of peace and love to counteract the fear. Even if we have reason to be fearful that the world will end in twelve years, we have to understand the power that we have. We don't have to accept anything. We can take responsibility for the earth, and we can change it.

THE AURA TECHNIQUE

How can you change your reality with your mind and actions? I'm not talking about a little change, like a new pair of shoes, I'm talking about big, miraculous changes, such as a massive surge of wealth and the type of unbelievable health and happiness that make it feel as if every day is a holiday.

What do you turn to when you're trying to overcome a problem and are struggling to imagine how? This isn't about little, day-to-day things, but the big changes and things you really want to accomplish—what do you do when you've tried everything and nothing seems to work?

You've gone to specialists. You've read books. You've done all you can, yet you can't seem to break through. I often ask myself whether there's a possibility that already exists in the quantum field—a solution, a mechanism, or a way to move into a different reality. If I don't know the answer, and many times we don't, the next step is to move into the unknown. It's like planting a seed. You wouldn't plant two trees next to each other, because they need space. Sometimes we need a little space for the unknown as well.

Every day, we wake up to the same memory of experiences and reaffirm the same idea of the future. As long as we follow these routines, our brains become a record of everything that we've known, all influenced by the environment around us. The only way to break out of the neurological structures that force us into our environments and memories of the past is to actualize a different reality.

This is the process that I have honed and taught to many people, with profound results. I call it the AURA technique. The AURA technique has two simple steps. The first step is alternate universe reality activation, and the second step is alternative universe reality actualization. The path between these steps is key to making those big changes in your life. There is a reality that exists right now in an infinite information field, but first you have to find and activate it. Then you have to actualize it into material reality. The intention is to find that parallel universe that you want, to activate it, and to go about the process of actualizing it into the material reality of your intention. All of this starts by first acknowledging the neurological structures that we're trying to break out of.

UNDERSTANDING THE HEART AND MIND

Biologically, we have two personalities: the brain and mind, and the heart and body. The brain is a memory bank. It's an archive of everything you have experienced up to that moment. As we saw with habits in the last chapter, waking up in the morning activates circuits in the brain that are nothing more than memories of the past, and each memory has thoughts and emotions associated with it. Thoughts are the language of the brain, while feelings are the language of the body, and every day we speak the same languages over and over, which creates our reality. The biology of the past makes us crave a predictable future. The known is safe, even when we have goals that stretch beyond it.

Because we can't easily move ourselves beyond that predictability, thoughts and feelings become a reality hack. The way we think and feel—the languages of the brain and body—create a state of being. Together, they define a moment in time that's connected with our personality.

Neuroscientists have found that the memories we call up easiest are long-term memories connected with strong emotions. Those moments alter something inside us, capturing the brain's attention and diverting the brain's resources to take a snapshot of that moment and create an internal anchor.

Over time, as we think about that moment again and again, we talk about it to our friends, we pay attention to it and remember it, we start to create circuitry around it. We start to feel it and hardwire it into ourselves until the body believes it is real and present.

These feedback loops condition the body to literally live in the past. It can't tell the difference between the actual experience and the thoughts and emotions connected to it. In fact, it usually takes some big crisis event to wake us up out of the past to even observe the present. We have all had those moments where we reach our lowest level, then start paying attention to how we're acting. We often start to notice suffering that we hadn't realized before—some people look back to realize they had been suffering for twenty or thirty years without knowing it. Our minds compartmentalize the experience to say it is not suffering, while our bodies ignore the point of time we're in.

The key to breaking these feedback loops exists in the heart. Once your heart and mind are unified, you can begin to see miraculous changes in your material reality beyond anything that you can fathom. The heart is your connection to the infinite field of information that exists all around us. Through your heart, you can sense the flow of the future.

WAKING UP, STRETCHING, AND STARTING OVER

If we go back to the metaphor of living in a movie, it's like the body is stuck watching an archive of films, and those archives inform the way we subconsciously direct.

Our bodies are thinking for us. Waking up in the movie is becoming conscious of programming—becoming the observer, objectively watching yourself in this moment.

Although time is nonlinear from a quantum physics standpoint, most of us assume a much narrower view. We can only walk in the present or relive the past, because we can only look for the knowns. We're working from a place of survival, where opportunities in the future are too much to comprehend.

Using the AURA technique, you can shift into what I believe are alternate, parallel realities. This is a shock for most of our bodies at first. They have conditioned us to stay in the moment, where it seems safer—and sometimes, it is safer. If your reality is exactly what you want, that's great. But if you want to change where you are or to create your own reality, you can follow the exercise outlined in this book.

It could be that they really exist, or that they just exist in the information field. The mathematical formulation of it all is less important than use and practice. Don't get locked into trying to believe it. Instead, work to develop a practice. Find the will to become greater than your body and mind, and to recondition them into something new.

VIEWPOINTING

Writers and actors have a superpower that a lot of us do not—that is the ability to find other viewpoints. How do we identify someone else's perspective and then embrace it as our own? Before any other technique, this is a skill that we have to develop.

The very first exercise that I do before reality surfing is called viewpointing, and it allows us to take a view of who we are from the third person. We analyze ourselves separately, through the eyes of someone else, including all of the kinesthetic body feelings, thoughts, and deep embrace of someone else's perspective.

Can you take the viewpoint of people and things in your environment? For many, this is an unpracticed skill.

The viewpointing exercise extends beyond person to person, as well. You might imagine that the desk you're sitting at is conscious. What would its viewpoint be? What if the monitor you're sitting in front of every day had a consciousness? As you begin to enter the viewpoint of things all around you, you start to become aware of their energy. With practice, this becomes a muscle that you can stretch and use more and more over time. Start taking on the viewpoint of everything around you, including people, animals, and inanimate objects. By doing this, you will start to connect and control everything in your environment and make your visualizations more intense.

We do this because, if you want to move into an alternate reality, you need to be able to step into a viewpoint that you might not be totally aware of. You have to embrace it, understand it, and feel it as though it were your own all along—because once you are there, it truly will be yours.

DEFINE WHAT YOU WANT

Once we are comfortable stretching our viewpoints and before we begin the process of activating a new reality, we first have to define our desired reality. I have found that a lot of people don't have an exact, defined reality that they desire. It's hard for many people to find clarity on the reality that they are in and the one that they want.

For some people, this is a long process. They might joke around with it for a while. Maybe they will dream about a vacation to Florida, then they'll see it on TV and it will be provided for them. Others can go more deeply.

If you don't know what you want, sometimes you need to define what you no longer want to experience, then look for its opposite. We can get hung up on this step when we think it's about finding a specific thing that we want. I find this is especially true with people who come to me for coaching. Usually someone who has the ability to pay for a coach is already fairly successful. Their struggle is in finding their actual purpose outside of the material things. This isn't about getting concert tickets or buying a car. The best way for it to work is to treat it like reality, and the best way to identify reality is to connect with the emotions around it.

Write down this reality, even if it feels vague right now. One powerful technique for doing this is to clarify your vision into a journey of ten different places. The journey method makes your definitions easier to integrate with feeling, which becomes easier to remember. In his book

Quantum Memory, Dominic O'Brien suggests you can solidify a memory into a long-term memory by breaking it up into a journey of ten different segments. In the same way that actors take over a role, you can imagine walking through ten locations or moments. This journey should extend beyond just a thing or a moment. It should be an entire existence that defines everything about you.

Write it down as a way to embrace it fully, then memorize it. Create it in your mind in the way that the recent BBC version of *Sherlock* depicts his mind palace as a place to access thoughts and details. You should be able to sink down into that visualization once you reach that state of silence—not just a thing that you see, but the version of yourself who is already experiencing that reality.

When you are able to access this level of reality change, it will have a ripple effect into everything in your life. Make this envisioned reality something that you embrace and connect with on an emotional level. For me, the related emotions are often a sense of relief, freedom, or joy. When you move into the next step—into the void of zero point—it's this joy or emotion that you'll hang on to.

CREATE A SENSE OF OWNERSHIP

Because we're changing our own realities, we have to consider what we could have done differently in our past, which is where radical responsibility comes in. Embrace the idea that you were responsible for every moment.

Often, it's nothing more than our reaction that made a difference.

Even now, I take responsibility for the home invasion. I could have created it on purpose because it awakened me. I remember, when we won the tickets to Hawaii, the kids who would later become burglars were outside bumping to their music as we loaded up the car. I remember waving to them and them looking me dead in the eye. The whole trip, I would tell my girlfriend, "I'm worried those kids are going to do something while the house is empty." Some part of me could have invited and created it so that I could move to this reality and moment in time.

That doesn't have to be the full explanation of what happened, but it allows me to take some responsibility and control of my reality. Even if you're not responsible, the exercise empowers you to take some control back. If you could go back and change your reaction to that one particular moment, then play out a potential future from there, you have something to work with.

GUIDED VISUALIZATION

Picture yourself sleeping as the sun comes up. Someone is next to you—who is that in your bed? What kind of room do you have? What does it look like, smell like, sound like? Is the room clean? What kind of windows does it have?

When you wake up, where is your wallet resting? What's in it?

You walk over to the bathroom—what kind of space is it? What does your sink and bathtub look like. What do you see when you look up into the mirror?

When you're done brushing your teeth, you wash your hands and walk through the hall—where do you go? Do you have an office, a workout room, a pool, a theater room? Where are your children?

What are you excited about for the day to come? Are you about to travel? To work?

You walk to the garage, get into your car—what kind of car is it? What does it sound like?

You drive through the ATM on the way to your destination. What does it feel like to withdraw money?

Try not to create opposing forces. Complicating this can wither your frequency. Link as much of it together like a journey as you can. You won't call back to all of this when you're in zero point, but having completed this exercise will give you a complete, specific emotion to tap into.

ENTER THE VOID

Next, we go into zero point. This is a point often found in deep meditation, during which we are without thought. We can call this the source, the void, oneness, and zero point. In any case, it is complete silence, and it is completely necessary. Before receiving something new, we have to return to nothingness. We have to access our operating systems with uninhibited minds.

TOOLS FOR MEDITATING INTO THE VOID

I have designed a meditation to help with this process. To help you go into zero point, try the void meditation.[20]

For a more complete discussion of this concept, check my *Reality Revolution* podcast episodes about accessing source through the void and understanding the space of variations.[21][22]

---

20 Brian Scott, "Guided Meditation: The Void Meditation," Podcast Audio. *The Reality Revolution Podcast.* May 2, 2019. http://www.therealityrevolution.com/guided-mediation-the-void-meditation-ep-15/.

21 Brian Scott. "Accessing Source through the Void to Achieve Your Dreams and Create Your Reality." Podcast Audio. *The Reality Revolution Podcast.* March 23, 2019. http://www.therealityrevolution.com/accessing-source-through-the-void-to-achieve-your-dreams-and-create-your-reality-ep13/.

22 Brian Scott. "Understanding the Space of Variations." Podcast Audio. *The Reality Revolution Podcast.* August 9, 2019. http://www.therealityrevolution.com/understanding-the-space-of-variations-ep-88/.

If we're locked into our memories, then we're also locked into material things. We have cell phones, homes, work, our bodies, our gender, our beliefs—all of these things combine to define what we have and what we do. They get stuck in our bodies, in the way we engage in sex, in the way we survive, in the way we use willpower. Once you reach the heart, however, everything changes. That's when, in a sense, we shift from particles to waves.

When we talk about accessing the source, it's the space where anything is possible. It's infinite emptiness with no one and nothing around, just pure possibility of thought. Many people call it God or their higher self. We know it as a unified field that causes all vibrating matter to come together at the speed of light. It's where particles become entangled—where one that is destroyed causes the other to be destroyed light-years away. It is where oneness happens, the link between all of these subatomic particles. A literal space that we can access. This is the space that's referred to in reality transurfing as the space of variations.

Humans have attempted to access this space for ages upon ages. It is where meditative practices and flow states are found. To change our reality, we have to venture into this void. I have used everything from Dr. Joe's hypnosis techniques to meditation to reach this state, though getting there is a lifelong goal. Once you reach it, doors open that you never imagined opening. It's exciting, and it draws you in for more.

Meditation often sounds complicated, but in this process we simply need to find zero point. Ignore issues or problems. Ignore factual evidence of things happening around you. You can't bring any of yourself or any of your baggage. You can't bring your identity or your history.

Often, we talk about using imagination. If you went back in time and changed something in your past, then moved through the future that flowed out from it, what, who, or where would you be now? Without getting to zero point, you can't really let go of who you are now to get to who you could be.

The changes don't have to be heavy boulders, either. If a parental relationship has been hurtful, for example, it's not the parent that we want to eliminate—which of course brings up the concern of never having existed at all. Instead, there is something about the relationship with that parent that we want to change. To identify that dynamic, we can often go to a specific moment that could have changed things.

This exercise is similar to visualization, but more like a relaxed ability to tune into something that's already available, like an incredibly detailed, twenty-novel, choose-your-own-ending fantasy book universe. It has every detail that you've ever imagined, including information about the world you're in now and worlds you've never seen. At zero point, you can step into the viewpoint of any of those novels, not just as a mental event but with a complete sensory experience—touch, sound, smell, and emotions.

As you reach that state, you might feel a smile on your face. You might feel your heart rate jumping and your cheeks flush. Your breathing changes. Rest in that sensation for a few minutes. Notice whether there is a vibration or rhythm to that particular experience. Then as you release the sensation, you're entering a new point. You're resting into it, not forcing it.

FORM A PRACTICE

The AURA process is intended to bring us to a place where the solutions are. You might practice these steps regularly—and because of the nature of our technological reality, I believe a regular practice helps them sink in. You could practice these meditations for thirty minutes in the morning and again as you go to sleep. You might be able to tune into the void first thing without any meditation at all.

There is a flexibility to this practice that can adapt to your reality. The more you feel it, the less you need to do it. At some point, you'll be in that reality without any need for the meditation. Most people find that doing the meditation at least once a day, for twenty-one to thirty days in a row, is linked with larger success.

As a final word of caution on this method, as with all parts of this book, don't take it lightly. This is an incredibly powerful activity. Come to it from a place deep in your heart, not just experimenting or looking for vengeance.

This isn't a way to express your anger and fear, and if you can, you'll be pulled into realities of anger and fear where the worst things are happening. You have to prune the vines of thoughts in your mind before you become entangled in something dangerous. Don't react to your past or present reality. You're no longer in the same energy once you're done.

If you decide to get ridiculous and test me out and go envision yourself as the president of the United States, what might happen? Imagine a train that's moving very fast, and you decide to jump on that train. What's going to happen? It won't be the same as catching a slow-moving train. You couldn't be consistent enough to match the speed and jump on it, and it would likely kill you.

Maybe one day you are going to be president. But it won't be from testing me out. It would come from harnessing energy close to that point, being willing to take a long journey, and coming naturally into the viewpoint of that reality.

AURA is like surfing lessons. You start out on the board, just lying down until it's time. Then when you see a wave that you want, you use core muscles to pull yourself up to land on both feet. Then you keep your balance to stay on the wave. The first step is finding the wave and the second step is propelling yourself forward and riding the wave.

When my family and I took surfing lessons, my kids were free and relaxed, and they caught wave after wave.

My balance, on the other hand, was terrible. I fell so many times. It took a lot of time and balance to spot the waves I wanted to learn how to catch them.

We can learn to surf our realities, or we can run into the side of a train.

If you can reach that place of freedom and balance, where you're not supporting or against anything—you're a true observer—the stories I have heard are incredible. This can open up doors that you never thought were possible, in ways that reaffirm the incredible brilliance of the universe that we are in. The reality you envision will come to you in ways that you least expect and that leave you no doubt that it comes from a connection to something greater. It will inspire you to do this again and again.

Simply understand that all possibilities exist in the eternal now. Each problem has a resolution that already exists, and you are outside of both the problem and the solution. You are evolving out of it—beyond our Newtonian, defined, particle universe into a new matter wave where all of the possibilities are defined. In the same way that your body follows your subconscious mind to the bathroom in the morning, if you keep putting awareness into the future throughout each present moment, your body will follow your mind to this unknown place as well.

GUIDED VISUALIZATION

Imagine you're sitting in the middle of a void between all of the universes, with access to all of them. Sitting there in that bubble, you can focus on the frequency of the particular reality that you want. It may look like a dot. It may look like something specific standing out amongst all of the crazy colors in your mind. It may be in front of you or behind you—doesn't matter, you can see all around you in that state of unconsciousness.

As you follow that one point, it becomes bigger. You move toward it, and it comes closer and grows even more. You try to move into it. You explore it. You become it.

You don't bring anything of your own to it—you simply let it be what it is.

Often, the new reality is surprising. There are things about it that you don't yet know. You're okay with that. You rest in it. You get used to the feeling of it. You believe completely that it has been fulfilled. You know that there will be moments to come that test your reality, because your body wants to return to what is familiar. Maybe you moved into a reality where you're thriving in business, but three or four days later you lose your job. Fear will pull you away from what you intended, if you let it. A ten-second response can derail everything you've worked toward, so you choose to continue resting rather than feeding the pendulum of fear. You choose to wait.

You don't affirm or visualize or repeat. You don't hope for some distant future. You have claimed this version of yourself as the real you. There is no desperate sense of importance. You don't ask when, how, or where evidence of your new reality will show up. Instead, you simply do what is offered to you throughout the day, feel the body's sense of your chosen reality, and rest in it. Soon, a better job comes along, and the feelings that you envisioned are part of your everyday life.

PART II

TAKE CONTROL OF YOUR REALITY

CHAPTER FOUR

HACK YOUR SUBCONSCIOUS

"Know that in your deeper mind are infinite intelligence and infinite power."

—JOSEPH MURPHY

My first summer in college, I worked on a gigantic ranch in Wyoming. I had never ridden a horse before, and my coworkers joked that I should go ride the horse they were trying to tame. However, I didn't know that it was a joke, so I got up on the horse only to be thrown off several times. Once I understood what was happening and attuned myself with the horse, it calmed down. Eventually, I was even able to get on the saddle and ride the horse.

Something similar is constantly happening in our lives. Reality is a wild, untamed horse that's bucking and kicking us around. If we ignore that truth, we're just going to be bucked off. But if we can tune into reality through our subconscious mind and come to grips with how it all works, we actually gain control and can ride the horse.

Whether we ignore it, downplay it, or work with it, we are all aware of the power of the subconscious. It is one of the most important forces affecting everything that we do. The subconscious mind operates our automatic bodily processes like growth and repair—if you cut yourself it will provide blood cells and tissue to that cut until it is healed. It keeps our hearts beating, thousands of times every day, and manages hundreds of complex bodily processes, as well as our minds. One of the best books about the Law of Attraction came from Joseph Murphy's *The Power of the Subconscious Mind*, in which he outlines the incredible power of the subconscious mind.

The subconscious part of the brain is designed to avoid unnecessary energy consumption, since the brain itself requires almost 20 percent of the glucose and oxygen in the blood. In a conservation effort, the prefrontal cortex doesn't use any energy to discover the genesis of new ideas. Meanwhile, while we think that the decisions that we have made come from the "awake," conscious part of the brain, often we are making a decision that has already been made by the subconscious. This has been observed on EEG patterns, when the brain began to answer questions before the person had answered it.

This quiet but active part of the brain can be broken down into two functions: there is the computer subconscious that operates without us having any access to it at all, and then there is the part that is somewhat self-aware. It can be considered your higher self or your heart, but it has some level of personality compared with the automatic functions. The automatic functions can then be broken down into physical actions such as riding a bike or driving a car, and mental actions like attitudes. The subconscious mind houses memory, including real and imagined life experiences. While it can process unfathomable amounts of information at once—some studies suggest four hundred billion bits of information per second—it cannot edit that incoming information as true or false, helpful or harmful. It can, on the other hand, cure diseases and create and alleviate stress.

To put that into perspective, a movie has about twenty-four frames per second, while we're living in about nine million frames per second, and the subconscious mind processes exponentially more than that. The conscious mind is much slower, more deliberate, and less efficient—processing only two thousand bits of information, requiring them to be grouped, and moving its impulses at about 150 miles per hour compared with the subconscious mind's hundred thousand miles per hour. This makes the subconscious one of the most advanced forms of technology on the planet.

And this is the technology that we're trying to hack.

There are three-hundred-trillion-to-one odds of you being alive in this time. You are a living, breathing miracle. How hard is it to believe that a few dozen, or a few hundred more impossible things could happen in your life? The world has been conspiring for you forever. Unexpected doors are going to open. Let them. You would do anything for your family or your kids—do this for you, too. Believe that you deserve it. Know that it's coming and be grateful.

Your purpose starts today, not tomorrow. Accept that you are extraordinary and define what you want. The person you will be in five years depends on the actions you take now, the information you take in, the workouts you do, the food you eat, the things you think. Your future self is begging you to show some discipline. The actions you take now will equal your thoughts later.

Incredible things will manifest in your life, but they will happen in the seconds and moments, all along the way. The journey is what counts.

The subconscious is also the heart that guides us. It's the part that will bring us prosperity and grant wealth and intuition. It will grant access to past riches and relationships. It absolutely is the key, but it's tough. To transform our lives, we have to transform our thinking. To transform thinking, we have to enter and transform our subconscious minds. There is no other way to change your life—to make your future self proud.

Learning how to hack your subconscious mind is one of the key elements to reality creation. Here's what that might look like in practice: A client who came to me for coaching was extremely overweight due to unhealthy eating habits. This woman shared with me that she had been in several toxic relationships and was now afraid to find love because she linked relationships to pain. The two of us came to realize that she was eating to avoid what she imagined as the pain of finding love. Once we changed this woman's beliefs, she began to lose weight very quickly. Today, she is in a fantastic relationship.

While the details vary, the underlying story is common in the people that I coach. Usually, there is something happening on a subconscious level—some memory or belief holds them to a script. All of us are just following the program. Even if we consciously want to maneuver through realities in meditation, the subconscious mind will pull us back if we don't address it.

Whenever we're playing with reality, we're playing with the subconscious mind.

Often, this is a lifelong journey—the discovery that we can continually control the subconscious mind throughout our lives. In this space where our collected behavior is leading us toward mass extinction events, we need to harness these abilities for the sake of ourselves and our civilization.

THE MAGIC TIME

The mind doesn't know the difference between what's real and what's just vividly imagined. Through a distinctive form of guided meditation or subliminal tapes, hypnosis can override limiting programs that your subconscious has retained. About five minutes prior to sleep and waking, when we're already in that theta wave state, is a great time to access that space with or without hypnosis.

The classic movie *Ladyhawke* comes to mind, when Michelle Pfeiffer and Rutger Hauer have been cursed to turn into animals during the day—one into a hawk and one into a wolf. They are in love with each other, but can only see each other during dusk, just as the sun goes down. That's how our subconscious is available to us—in that beautiful moment just before we go to sleep or wake up. If we're aware of that period of time, we can take advantage of it and use that time to make contact with the subconscious.

ALTERING MEDIA CONSUMPTION

We can also bring ourselves into a feelings space by utilizing repetition of emotions throughout the day to create that sense of imagined, envisioned reality. Whatever we watch and listen to will inform our subconscious, like it or not. When we wake up and blare the news, listen to depressing songs, read dark articles, scroll through an awful Facebook feed, and constantly repeat millions of seconds of negative information, it's no wonder our minds say we don't have control, the world's crazy, and there's no hope.

Try a thirty-day experiment, where you actively choose every single thing that you listen to and watch every single day. Choose positive songs, uplifting shows, and positive books. If you want to go to sleep with the TV on, make it HGTV instead of the news. If you think that you love sad songs but always feel sad, change to something positive for a month. It might take some time, but this is one of the first things we have control over, and you'll likely notice a difference by the end of the month.

Listening to hypnotic or affirming programs while you sleep is a powerful addition to this experiment, although some people find that their partners don't enjoy it. I recommend a sleep-sound mask with a flat earphone that you can lay down on, that runs while you sleep or connects to your phone to play as you fall asleep.[23]

THE POWER OF AFFIRMATIONS

In the same vein, repeating affirmations on a regular basis and embracing the feelings around them will allow some of that subconscious programming to start to change in a positive way. I've met people who think it's absolutely ridiculous to repeat affirmations out loud. Others live in an environment where they don't feel comfortable doing

---

23 Try this playlist of sleep programming meditations: https://www.youtube.com/playlist?list=PLKv1KCSKwOo_5Sv8NSXuDWudAVmoDns6Z.

them. If it becomes an intention, however, you can always find time and space for them.

Simply repeating affirmations will only have limited benefits; you have to actually *feel* the affirmations. Monitor your feelings and thoughts as you use them. If you feel separate from the affirmation, you may project their opposite. For peak effectiveness, create believable and emotional affirmations.

You can do affirmations when you're in the car, walking outside, or even in the shower. You can also find thirty minutes of affirmations to listen to while you work out or fall asleep. Listen to them at normal speed and quietly say them to yourself, or turn them up to three-times speed and let them run over and over again. Give it a try and see how it goes. For most people, five minutes of affirmations is an incredibly long time—I recommend stretching yourself even further. Experiment with a full hour and see what happens. I love to start affirmations at the beginning of a hike. By the time the hike is over, the things I repeated are ingrained in the neurons of my brain.

Become an affirmation collector. If something resonates with you, write it down. When I see that someone was using affirmations when they won the lottery or their business did well, I will grab it up like it's a golden nugget and begin to repeat it with all of my other affirmations. You can also create your own, emotion-based affirmations that speak to you, but be careful what you choose to repeat and focus on. I once asked for large amounts of money, and it came to me in the form of a loan that I had to pay back very quickly. This taught me to be more specific.

If you want money, it should be money that you get to keep that is in the best interests of everyone involved. That might look like saying, "Large amounts come to me easily and quickly, in increasing quantities, from multiple sources, on a continuous basis, in the best interests of all, I gratefully keep with happiness."

Use the present tense and do not use words like "can't" or "won't." For example, instead of saying, "I won't be poor" say "I am wealthy."

Another powerful affirmation technique is listening to double-induction affirmations, which are affirmations played separately in each ear. This form of affirmations is based on a classic induction technique in which therapists speak to a client in both ears simultaneously. This forces the client's attention to go on one affirmation, while the second affirmation goes directly into the subconscious. For a good example, listen to my wealth and abundance affirmations, which use this technique.[24]

MORE SUBCONSCIOUS HACKS

We are constantly exposed to messages, even in our work-

---

24 Brian Scott. "500+ Financial Wealth and Abundance Affirmations with Binaural Trance Induction (DELTA WAVE)." Podcast Audio. *The Reality Revolution Podcast*. June 16, 2019. http://www.therealityrevolution.com/500-financial-wealth-and-abundance-affirmations-with-binaural-trance-induction-delta-wave-500-financial-wealth-and-abundance-affirmations-with-binaural-trance-induction-delta-wave-ep63/

place. There are resources we can use to help mold our subconscious, even while we are working. Belief is the beginning of all of this, and if you believe it might work, then try it.

I love to use a program called Mind Flasher while I'm working, which flashes affirmations across my screen at regular intervals.[25] There are several other similar programs available. These flashes occur in the matter of a millisecond; just long enough for your subconscious to register it. You can adjust the settings so that the affirmations filter onto your screen in time increments of your choosing. Imagine: if an affirmation comes up every five minutes over the course of an eight-hour workday, you have accessed affirmations ninety-six times in a single day. They can flash for just a millisecond, which is just long enough for your subconscious to register it.

You have the ability to customize your affirmations. For example, I once set mine to read, "I made $35,000 this month." Two months later, my net profit was—you guessed it!—$35,000 per month. I've had clients who were struggling to find a relationship use pictures of women they found attractive as a form of affirmation. Next thing they knew, they met a woman who looked uncannily similar.

You can use something like this as a vision board for the car or house or trip you want. You can make positive statements about yourself. You can use it for whatever

---

[25] Link to product: https://mc2method.org/subliminal-software

message you need your subconscious to understand—arguably, to a greater effect than affirmations have. It's one of the greatest, simplest inventions ever made.

Another way to stretch your subconscious and begin to learn on another level is to listen to books and YouTube videos at top speed. You might not completely understand the conversations at first, but you will find that you recalibrate over time. One trick I use is to speed up to two-times normal speed first, then dial it back to one-and-a-half speed after a couple of minutes, which will make one-and-a-half speed sound slower and easier to understand than it would if you were to adjust from normal rate up to one-and-a-half. Set goals and try to keep increasing until you're at three- or four-times normal speed.

The same goes for speed reading text. Move your fingers across the page to digest multiple words at a time instead of just one. Think of two words as one, then four, then a sentence, then a paragraph digested as a single word. It's difficult, but these techniques allow you to consume entire books in a day, and to ultimately retain and access the information from them easier as well. You might not understand it at first, but your subconscious can, and you can train your brain to keep up.

Not only can you consume more information in those contexts, but normal time will seem much slower. People will seem to talk with spaces in between what they're saying, and the subconscious will have more access in those gaps. You become more present and able to wake yourself up to break old scripts and subconscious influences. It gives you greater focus, and whatever you focus on, you get.

EMOTION AND ENERGY

Repetition alone is powerful, but it really needs emotion to round out its effectiveness. Someone with cancer saying "I'm healthy" is not just going to work. But if you get in touch with the feelings around it and repeat that emotion over and over, there's a point in time where your subconscious almost says, "Whatever you say! I know it's not true, but okay, enough already." Bringing feeling into the repetition breaks down lack of belief.

The more we learn about the subconscious mind, the more we realize we are facing an epidemic of trapped emotions. It's easy enough to express positive emotions—when we're happy, we tell our friends, our parents, and our kids. But when we have fears, stress, and terrible situations happen, those emotions stay held inside of our bodies. These emotions become energy that manifests itself by pulling us into realities where we face disease and repeated illnesses.

The subconscious mind is directly involved in bodily processes, particularly in muscle memory. According to *Subconscious Mind Power* by Kevin L. Michel, most of the time, sickness in the body originates from something mental. According to Jonathan Tripodi, whose thoughts are supported by leading-edge researchers and his own discoveries, properties exist for information to be communicated and stored within and among every single cell of the body.[26] There is some question, for example, around people who drink excessive amounts of alcohol in connection with their anger. When their livers are damaged, was it caused by the alcohol or the anger? Stored negativity might come from loss, financial problems, relationship problems, work stress, abuse, combat trauma, rejection, negative beliefs, and negative self-talk. The emerging idea is that, when we unlock stored emotions, we can free the part of the body that it affects.

Dr. Robert Bradley Nelson says, "If you have a trapped emotion, you will attract more of the emotion into your life. You will also tend to feel that more readily and often, more often than you otherwise would." Without releasing emotions, we're pulled into realities where we feel more of them. Intentional release gives us the freedom to move into better states of awareness and the realities we'd like to manifest.

> When suppressed anger is released, you might become hot, clench your fists, or scream. Suppressed sadness might look like lung contraction while tears fall. To intentionally release negative emotions, you have to be able to communicate with

---

[26] For more information on this, I recommend reading *Freedom from Body Memory* by Jonathan Tripodi, *The Emotion Code* by Dr. Robert Bradley Nelson, *Awakening to the Secret Code of Your Mind* by Darren Weissman, *The Secret Language of Your Body*, by Inna Segal, and *Subconscious Mind Power* by Kevin L. Michel.

> your body, so there are physical exercises that you can do to trigger the body's natural reaction while you ask your subconscious questions. Dr. Robert Bradley Nelson's book has a matrix that identifies an emotion and the part of the body it connects to, with a process to release them.

Another step is to harness energy psychology, which is a way to change our beliefs, often to provoke super learning or to achieve something specific. Willpower is a great example. Yes, it is the ability to take an action that's outside of your comfort zone, but more than that, it's an energy. Understanding it as an energy that is directly connected to the subconscious helps us to harness it better. For example, when we know that willpower declines just as any other energy does, we can give it more energy by avoiding temptations and developing self-control.

A big part of energy psychology includes reaching subconscious muscle memory through body movement or synchronizing the hemispheres of our brain to rapidly download something into the subconscious. Qigong, emotional tapping along the meridians of the face, tapping on the chest, emotional freedom techniques, data healing, and holosync are all used in this capacity. For more information about how to hack your emotions, you can check out *The Reality Revolution Podcast* on this topic.[27]

---

27 Brian Scott. "Hacking Your Emotions." Podcast Audio. *The Reality Revolution Podcast*. March 23, 2019. http://www.therealityrevolution.com/hacking-your-emotions-ep8/.

POSITIVE THINKING

Priming exercises in the morning set the tone for the day. Writing out grateful statements about the present or the past remind you to be positive and grateful and to hold onto that mindset all day long. This creates a frame, like the way a movie scene is framed and filtered. The frame and filter of gratitude gives you permission to do the things you want to do. You'll make better choices and feel better about them as well.

Another positive shift is to become compassionate about your friends and family. Create moments in time where you pray for them, to think about their needs. These moments take us away from ourselves and move our focus to others. When we let the subconscious think about itself, we begin to lose energy. Outward focus on other people on the level of compassion and care brings that energy back and begins to change the way the subconscious thinks and acts in the first place.

Radical forgiveness is linked with both of these concepts. Many of us reach times in our lives where we absolutely hate someone. They might have done something absolutely terrible, and we feel like we can't forgive them. In my case, someone tried to kill me.

If you can come to a place of radical forgiveness, it will unlock your subconscious mind to do powerful things. Harboring unforgiveness is like a jail of anger and hostility that holds our minds hostage. It becomes a sphere that locks in all potential. In *The Power of Your Subcon-*

scious Mind, Joseph Murphy says that if you think about someone and a bit of anger arises, you have not forgiven them.[28]

Create a statement or phrase that outright forgives them and wishes them the best. Each time you think of them, release them with that statement. You aren't doing this exercise for them, but for yourself. With that in mind, don't forget to forgive yourself. I believe we can change our actions from the past but if you don't believe that, it can be difficult to reconcile who you are now with who you were then. You can choose to live in the past and become addicted to the anger and resentment you have, or you can open yourself up to a new level of emotion that can transform your life from the subconscious level.

A client of mine committed armed robbery when he was young, and when he came to me, we found he would always stop himself before any kind of emotional breakthrough. He felt so guilty about what he had done, even though it happened twenty years before. I explained to him, like Joseph Murphy says, that the person you were all those years ago is literally not you. All of your body's cells have since been recreated. You are a different person. If you forgive yourself and radically accept yourself, then you are forgiven no matter what you have done. This can unlock the ability for you to do even greater things that will outshine and overcome anything you have done in the past. Once my client let go of his past and reduced the

---

[28] Joseph Murray, *The Power of Your Subconscious Mind*. (Martino Publishing, 2001).

importance of his past actions on his present life, his life began to change dramatically. He found a new job and entered into a great new relationship.

HIGH-IMPACT EVENTS

If you were to go back through your life to look at the things you remember, each of those memories will be of high impact events. An apparent crisis, a critical illness, a loss of a loved one, a major emergency, a significant blessing, the birth of a child, meeting a soulmate—these call for a complete departure from our normal ways of being. In these moments, a powerful window for holistic change opens up. Events like this create a neurological signature that is impressed upon your long-term memory. In fact, extensive psychological research confirms that events that happen during heightened states of emotion, such as fear, anger, and joy, are far more memorable than more mundane occurrences. Multiple studies clearly show that amplified states of emotion facilitate learning and memory. This happens because the parts of the brain where memories are stored need to distinguish between significant experiences and those that carry less importance. Based on this information, our long-term memory gives priority to the experiences that occur during heightened states.

The idea, then, is that we can create high-impact events and open that window by creating emotions around them and visualizing them happening. If you can create a high-impact event as though it were a memory—not unlike the emotion-attached visualization that Law of Attraction books talk about—you can actually pierce the bubble of the subconscious mind. Visualization functions as an experience for your subconscious, which stimulates neural chemicals and enhances the structural growth of synapses. In other words, you can create the circumstances that make something memorable without actually experiencing them directly.

Additionally, if the quantum field carries an informational signature with all possible realities available to you, and if the heart can read the emotions from this field, it will be easier to identify events associated with heightened states of emotion. Perhaps your heart can hear this future even better. By tuning into powerful emotions and feelings of joy and gratitude in your visualizations and meditation, you may be drawing future events to you that create these emotions.

WORKING WITH DREAMS

Everything in this book, from changing the subconscious to reality transurfing and parallel realities, connects back to our dreams. We each spend a third of our lifetimes sleeping, and scientific research only offers a limited understanding of what's happening in that time. We do know that, while we are dreaming, the subconscious mind is likely using those dreams to resolve issues and face problems. To take that a step deeper, if the brain is a multidimensional interface as we explored in chapter 2, then we may be able to access these parallel realities through our dreams. In any case, becoming a lucid dreamer is one way to influence the subconscious and thereby change your life. In all of my research and experience, I maintain that this is one of the key things to learn and practice.

Think about the dreams that you remember. Usually, they follow a script that you have not consciously written. In the dream, it often feels like watching a movie, being pulled along without actively making decisions. What we don't realize is that the same thing is happening when we're awake. We're following scripts that stretch all the way to the end of the timeline that we're on. The idea, then, is to become lucid in a dream state in order to become lucid while awake.

This is a huge hobby space, because once you can accomplish it, lucid dreaming becomes the ultimate entertainment. Andrew Holecek says, "Your mind becomes the theater, and you are the producer, director, writer, and main actor. You can script the perfect love story or the craziest adventure. Lucid dreaming can also be used to solve problems, rehearse situations, and work through psychological issues."[29] There are books and conventions and gatherings all centered around lucid

---

29   Andrew Holecek and Stephen LaBerge, PhD, *Dream Yoga*. (Sounds True, July 2016).

dreaming. It includes three stages: strong motivation or intention, dream recall, and an induction technique.

To begin, set your intention by setting alarms that chime every hour that you're awake during the day. Each time, ask yourself, "Am I dreaming?" Of course you know you're awake, and at first it will feel crazy to ask yourself this question. It doesn't matter. Eventually, you can let the chimes go off at night as well. Andrew Holecek recommends using dream signs alongside timers. All day and night, whenever something out of the ordinary happens, ask yourself whether you're dreaming. Use events as triggers to conduct state checks. When you dream, it's usually about things that you commonly do while awake. If you make it a habit to ask yourself whether you're dreaming regularly or when something odd happens, you'll begin to ask yourself the same question while you're dreaming and will start to gain a little bit of control.

By doing this, you are also waking yourself up to the present moment. This meta awareness is a very powerful place to become proactive about creating your reality. It is in those moments that you awaken yourself that you can become aware of the flow of reality and choose different scripts and activate new realities in the quantum field. This habit of lucidity checks has empowered my personal reality creation experiments immensely. Vadim Zeland discusses this beautifully in his book *Tufti the Priestess*, in which he writes that the first step to reality shifting is to wake yourself up and focus your attention on your awareness center. The awareness center is that part of you that exists between your inner and outer self. From here, you can simultaneously observe your thoughts and what is happening around you. You can see your surrounding reality and yourself within that reality.

If you are willing to disrupt your sleep cycle, another technique you might try is to wake yourself up approximately every three hours. If you can wake yourself up and go back to sleep, you'll stay pretty close to that dream state. As you fall back asleep, become aware of how much you want to dream. Set that motivation and intention. Just a word of warning that sleep is important. Regularly breaking up your sleep cycle reduces your energy, so don't make this a regular habit.

When you begin to dream, if you have any small moment of lucidity, try to look at your hand or into a nearby mirror. These are strange, seemingly insignificant things that we don't usually do in a dream. If you can make it happen, then it creates a fairly strong sense of self-awareness.

Another thing you can try is to do something completely crazy within the physics of that world. Once you know that you're dreaming but you're still asleep, transform the table into a dog. Go outside and try to fly. Do anything that you know you would not be able to do in the normal world.

As an example, a common dream of mine is that I've been sucked into a maze in a house where I cannot find my way out. When I become lucid, I fly right out of the house through the roof, breaking all of the walls like Superman, then look out over the sunset and fly off. Or in the dream that I have not finished my paper for school and I'm going to get a bad grade if I don't finish it. Once, in an early moment of lucidity, I snapped my fingers and the paper was done. In the dream that you're naked talking in front of a bunch of people, magically make your clothes appear. Do things that are outside of the norm to confirm the dream state and your control over it.

Once you have found yourself in a lucid dream, ask your subconscious mind questions. Ask what the next chapter is in your book. Ask how you can solve a specific problem. Ask what you're facing next and how to handle it. Then figure out how to keep that information. I like to write the dream down in a book that stays next to my bed. Dream recall is a difficult step, because we often lose the information in that short period of memory that disappears after we're fully awake. Have something to write on in a place where you cannot miss it—next to your bed, by the sink where you brush your teeth, by the coffee pot. Even before you reach lucidity, make a practice of writing down your dreams to improve recall, and it will improve your ability to control your dream states.

The *transurfing* model asserts that dreams may be unrealized sectors of the space of variations from the past and the future. You may be in another person's timeline, simply acting out the script of an unrealized future or past. By becoming lucid, you are more interactional in this space of variations and conditioned to living intentionally as you do in a dream. If you want to fly in a dream, you don't just wish to fly or pray to fly; you just fly.

Try out my lucid dreaming meditation, which will help you move from a delta wave to a theta wave. It allows you to awaken in the dream with subtle prompts that will occur several hours into your sleep cycle.[30]

> You can purchase a Remee sleep mask that has lights on it that light up in a pattern as you sleep.[31] When you dream, the lights become a signal from the outside that let you know you're dreaming. Dream goggles sense eye movement that indicates REM stage sleep before queuing lights. These are useful tools, but the key is to set the intention before bed that you want to have a lucid dream. Say it to yourself, write it down, and keep practicing even when it doesn't work right away.

You may be concerned about how this process might affect your sleep. The idea of a lucid dream seems like you won't get as much actual rest at night. In actuality, you're still experiencing normal restorative sleep. You might feel a little disrupted at first as you adapt to the

---

30 Brian Scott, "Guided Sleep Meditation: Lucid Dreaming – Become Lucid in Your Dreaming and Waking – Delta & Theta Waves." Podcast Audio. *The Reality Revolution Podcast*. August 31, 2019. http://www.therealityrevolution.com/guided-sleep-meditation-lucid-dreaming-become-lucid-in-your-dreaming-and-waking-delta-theta-waves-ep-105

31 Link to product: https://www.amazon.com/Remee-Lucid-Dreaming/dp/B0785JR4XM/

chimes or catch yourself waking up all the way—but you will get back to that point of eight, restful hours of sleep. You'll look forward to sleeping, too, because that is when you can literally live out your adventures, work out your problems, and understand different realities.

We do this because we are tired of living life passively while we're awake. Not only can we gain control in that lucid state during sleep, but this changes the way we think about our waking moments as well. I've found that asking myself if I'm awake during the day has turned into affirmations: *I'm awake. I'm outside of the script. I'm in this moment.* If you do something that feels like your subconscious took over—like you were following a program—you can "wake yourself up" and begin to take control.

Become aware of the script that you're in. Become awake enough to choose what you want to do—to choose the reality you want to have.

Dreaming in a lucid state allows us to begin to tune in to the power of intention. When you are dreaming, you can choose what you want and the environment around you will change to facilitate and accompany your desires. You can fly, sing at a rock concert, or spacewalk. The more lucid you become in your dreams, the more lucid you become while awake, which offers you the power to change the outside world in a similar way. While there is a delay when this is done while in a conscious state, tuning in to this intentional pattern is a game changer and will allow you to rule your reality.

RESCRIPTING

I'm the person who talks to myself in the shower. Until recently, I had never really stopped to ponder what I was actually saying to myself or thinking during these conversations. When it occurred to me to become more aware and write down my own words post-shower, I was in for a surprise. "You will never make it" and "There is no way" are just a couple of examples of the way in which I spoke to myself. While I was using positive affirmations and maintaining a positive mindset in my more conscious moments, I was mumbling negatively to myself unconsciously.

I knew I wanted to change this, so I tried a powerful technique that allowed me to monitor my thoughts. It works like this: each time I had a thought I would say it out loud then I would write it out. This is important because thoughts and unconscious actions can pull us into negative scripts. We don't want to forget these things we are saying or thinking—we want to gain control. Try writing down everything you think for a day and see what happens. When you speak or write out what's actually happening in your mind, you'll start to see patterns, and that awareness will allow you some control.

Another exercise to begin changing our scripts is to consciously *intend* to do everything.

Try this for a day or even for an hour. It's difficult to do, but decide to create an intention for every single action that you take. Intend to go to the bathroom, then get up and go to the bathroom. Intend to brush your teeth, then pick up the toothbrush and do it. If you can, visualize yourself doing the action as you set and say the intention. This will bring your subconscious in line with your reality. It's not easy to intend to go to the bathroom then grab coffee instead. When your subconscious begins moving in line with your intentions, it won't know the difference between mundane routines and crazy things that you want to accomplish. Give it a script to follow, and it will. I know this might sound easy, but I found it very difficult to do for even for a limited amount of time. Once I could go for a full hour, I felt like I could truly ride the horse.

This is not magic. The science behind all of this is called neuroplasticity—the idea that our brains actually change structure based on the way that we operate them. All of our executive functions and decision-making comes from the limbic area of the brain, which hardwires basic drives of good and bad, hunger and satisfaction, and momentary emotions. We want to break that subconscious decision-making down and reshape it into something we have more control over.

We want to understand and control the limbic area. We want to understand and control the hypothalamus and its cortisol releases that make us sick and broken down. We want to understand and control the amygdala, which is the fear-based part of the brain. We want to understand and control the hippocampus, which converts moment-to-moment experiences into memories.

We want to understand and control the frontal cortex, which allows us to think about thinking and deal with ideas and concepts. None of these functions are problematic or need to be turned off—they are how we survive in the world. However, we do want to become aware of them and to interact with the subconscious mind as it controls them.

Neuroplasticity—the idea that the brain shifts and changes, like a computer that can rewire itself—is a miracle. Whatever chemicals are being released, however your brain is functioning now, you can use thoughts to actively change its shape, structure, and function.

As we gain control over the subconscious, we're no longer subject to its whims. The secret to success is often the ability to do the emotionally difficult thing—to have the willpower to do the thing that you'd rather not do. To forgo one hundred dollars now for five hundred later. Yet most of us remain subject to our subconscious impulses. Most of us will choose immediate gratification.[32] Meanwhile, highly successful people are maneuvering through incredible realities because they have a sense of control. They do difficult tasks now that set them up for a future success. They face the emotional difficulties and actively shape their minds toward better health, meaningful friendships, deeper love, profound happiness, and greater prosperity.

---

32 W. Mischel, Ebbesen, E. B., & Raskoff Zeiss, A. "Cognitive and Attentional Mechanisms in Delay of Gratification." *Journal of Personality and Social Psychology*, Feb 21, 2 (1972): 204-18.

ACKNOWLEDGING THE TRUTH OF REALITY

If all probabilities exist in the quantum field, then there is a space in that void where all of the endings of our actions exist as well. That we'll die at sixty down one path and at a hundred down another. That the love of our lives can be found after this action and becoming a billionaire happens after another. Our brains are too rational to access and understand this trove of information, but our hearts—our subconscious minds—can. We might not remember it, but it can be processed alongside the other billions of bits of information that cycle through the subconscious every day.

When you go to a restaurant to eat, your mind took in all of the information about every person in there, every conversation within earshot. At a football game, it took in every sound, every face, every moment. There's a part of your mind dissecting every bit of information that it's exposed to every day, in every circumstance. It knows who wants to hire someone like you, who needs a leader like you, who knows where you should invest, and where you can make the connections you need for the next thing you want out of life. Your mind has that information.

If you're locked into your own path and who you say you are, you might ignore that incredible information that's available all around you. If you hang onto the kind of person, friend, worker you are—the gender, personality, and past that you have, and all of the little pieces of baggage that you carry around—you'll never see what's possible. Only when we drop the veil and connect with our subconscious can we access the source.

These exercises are not about manipulation, but about accessing and interfacing with our hearts. The subconscious has no concept of money or possessions or careers. It knows feelings and energy. It knows right and wrong. Step by step, we can connect with it, let go of old programming, and become more in tune with the present and our desired futures. It might look like spontaneous actions—your heart might tell you to take the left turn when home is to the right, or go to the store when you're supposed to be somewhere else. It's not just one big opportunity, but hundreds and hundreds of small ones.

Our hearts and minds are always moving toward whatever our attention is on. One second of a doubt can create or destroy everything. Sometimes, as soon as someone begins to come to grips with the power of their mind, they will immediately become destructive. They'll focus on the how and the why, instead of the impossible and the potential.

We have no idea what has already happened for you to even exist in the world. If you went back to trace all the things that could have happened—the people you didn't meet, the songs you didn't listen to, all of the intricacies of everyday life over all of our lifetimes—you would be blown away. The universe is infinite, and your life is infinite. Every choice you've ever made is infinite, and the results you can tap into are infinite, if that's what you truly want. We can see the tip of the iceberg, but a whole mountain of subconscious exists under the surface.

When we allow ourselves to connect with that iceberg—to change it and be changed by it—we can access incredible futures, where we achieve higher potentials beyond our wildest dreams.[33]

---

[33] For further exploration of how to hack the subconscious mind, check out this episode of *The Reality Revolution Podcast*: http://www.therealityrevolution.com/hacking-your-subconscious-mind-ep5/.

CHAPTER FIVE

WHY ENERGY MATTERS

Energy may be likened to the bending of a cross bow; decision, to the releasing of the trigger.

—SUN TZU

What do you think of when you hear the word *energy*? Do you think of its synonyms—vitality, vigor, life, animation? Defining energy is a difficult task, because it encompasses so much. It exists on a scientific level—the power directed from the utilization of physical and chemical resources. It exists on a mental level as the strength and vitality required for sustained physical and mental activity. In physics, it's the capacity for doing work—kinetic, thermal, electrical, chemical, and nuclear, all in various forms of potential and existence.

Energy is not easily defined, and is not always measurable. When we look at our bodies, our surroundings, and the physical world, we see solid, manifest forms—when in reality, on a subatomic level we are all just vibrating atoms of energy, moving fast enough to be perceived in our minds as solid and accessible.

When anything is accomplished, on any single level of life, it can be traced back to energy. It is the seed that allows reality itself to grow. Just as a stream of water can turn into ice, energy can manifest itself in different ways. Space, time, objects, nature, people, and planets—as well as the nonphysical universe such as thoughts, emotions, and various planes—are all just energy.

Likewise, everything in life has an energy signature. The vibrations exist on different frequencies, flowing in and out and up and down. Each variation in energy affects the way we maneuver through reality to manifest our dreams and goals. Just as the cells in our body entangle, they also seek other energy to connect with on a quantum resonance level. Without an entangled match—whenever the energy of people, places, and objects cannot attach to us because the resonance does not connect—we exist alongside each other as particle clouds.

There are two forms of energy in the human body: physiological energy and free energy. Physiological energy is attained via the digestive system. Free energy is the energy all around us that passes through the human body. Together, these two energies make up our energy body. Human energy radiates outside of our bodies and into the space around us. There is a never-ending supply of free energy in the world, but we only use a very tiny fraction of it.

Energy moves through our body in two directions. The first current passes in an upward direction. According to *Reality Transurfing*, in most cases, it runs one inch in front of the spine in men, and two inches in front of the spine in women. The second current of energy passes in a downward direction and runs very close to the spine. The amount of free energy a person has depends on the width of their central meridians. The wider a person's meridian, the greater their energy levels will be.

For this chapter, we will focus on these connections and the energy that flows between them.

CORDS AND LINKS: EVERYTHING IS CONNECTED

If you can imagine wires or hoses connecting us to people, places, and things all around us, sometimes those connections can deplete or recharge us. You've probably felt this in a conversation with someone where you felt completely drained and didn't know why, or if you've had a thought about someone who wasn't around. These could be signs of energetic connections, which exist all around us. I believe we have energetic connections to our past, our ancestors, our family, and our friends, as well as the items in your office and your home—everything has an energetic signature and can form a connection for the exchange of energy.

A natural attraction to someone or something, or an inherent agreement or likeness, might signify a close resemblance or connection that can have quantum resonance or coherence. The connections create Einstein's "spooky action," faster than the speed of light and at great distances. This is the entanglement we talked about in chapter 2, which makes particles behave as though they are the same even at great distances. If we become entangled with people, ideas, and pendulums, we are as affected by those things as much as our own heart, blood, and body—to the effect of positive or negative resonance. That is not to say "good" or "bad," but simply an energy increase or depletion.

These entanglements can be infinitesimally small, like microfilaments or spiderwebs, or they can be thick ropes of strong energy. They can be flexible, flowing, and soft, or they can be rigid and seem unmovable. We can envision them as iridescent and shimmering with light, or dense, dull, and murky. They are pathways to subconsciously send and receive information, not just etheric but otherwise meaningless connections. Highly intuitive people, such as shamans from Native cultures, have been able to sense these connections, sometimes as emanations from the solar plexus, third eye, groin, or chakras. Some people see the connections as colors. Others can see the links without realizing what they are seeing, because we are talking about literal links where energy is exchanged and maneuvered like we might navigate a highway or intersection.

Between two people, a relationship in either the positive or negative sense forms malleable energy filaments between you and that person. Information, emotions, and energy frequencies transfer between those links. When someone is resentful of another person, the connection becomes murky. When a group of people create a pendulum, our realities can be pulled into connection with it—such as a party, a group, or even a family. The more emotion involved in a relationship or the longer you've been in connection, the stronger the energetic attraction and bond will be. You might be tired one day because of some ex-boyfriend or girlfriend who is still connected to you in some way. Or you might find free energy being spent on excess importance, creating a palpable imbalance. If we're not at least aware of these connections, they can deplete us or draw us in where we don't actually want to be.

Underlying our discussion of energy flows in this chapter, there are three tenets to understand.

First, that everything is composed of constantly moving, changing energy. In spite of our common understanding of the universe as fixed, physicists acknowledge that all of life is and has been energy since the spawning of the world.

Second, that we are not separate from the world around us. In all of our technological advancements, we have forgotten this primordial wisdom. We are not solid, separate things existing in parallel. All creatures and concepts on this planet are inherently connected, constantly interacting with each other on quantum levels as well as at a larger scale. To extend this a step further: at one point in time, the particles inside of us could have been entangled all the way back to the beginning of time. All of the atoms and chemicals that exploded from our pea-sized universe in the Big Bang were connected—which means we may well be connected to the very stars we see in the sky. This is something that is difficult to write or speak about, but that we understand on a very deep, internal level.

Third, that everything has consciousness. Scientists are beginning to prove this—such as the understanding that plants have intent and respond to human energy fields. It may be on a minimal level, but everything from a pencil in your hand to the rock you use as a paperweight has some amount of consciousness. This is something that Native cultures have long recognized, while Western cultures have lost that understanding.

There's no specific trick to severing or creating links—no magic incantation to recite that will create an ideal level of connection with your surroundings or past. We're simply pulled into whatever we're looking for. If we intend to maneuver through our reality, we have to begin with an awareness of these billions of variations of energy and the way they coexist in connection with each other.

MEASURING ENERGY LEVELS

The way we interact with these cords of energy affects the reality that we're in. When you become aware of their presence, you can start to analyze each emotion and interaction in a brand-new light. Unfortunately, each thought can be its own energy, which quickly becomes difficult to identify and manage. The best scale for measurement that I've found was created by Frederick Dodson in *Levels of Energy*, based on David Hawkins' map of consciousness.

As we move up the scale, each level affects your mentality, the kind of relationships you have, the reality you create, and the abundance you create. Your consciousness changes along the scale.

None of the levels on the scale are permanent—you can have a level thirty day and a level eighty day, and you can move up and down on the scale from time to time. Just remember that frequencies are attracted to each other.

Here's what the scale I use looks like at some key points:

- 30: guilt, shame, psychosis, humiliation, hatred
- 50: apathy, despair, depression, hopelessness
- 80: grief, sorrow, self-pity
- 100: fear, worry, shyness, inferiority, paranoia
- 120: craving, need, compulsion, unfulfilled desire
- 160: anger domination, aggression, coldness

- 180: antagonism, criticism, content, complaint, blame
- 190: pride, superiority, arrogance
- 200: contentment, routine, functionality, boredom, low levels of energy
- 275: courage, relaxation, eagerness, fun
- 320: willingness, kindness, optimism
- 400: acceptance, attention, neutrality
- 450: intelligence, knowledge, reason
- 475: joy, creativity
- 505: beauty, creativity
- 510: power, initiative, integrity
- 530: love, intuition, appreciation
- 540: humor, happiness
- 550: unconditional love
- 570: ecstasy, exultation
- 600: bliss, peace, serenity
- 700: oneness
- 1,000: infinity

You don't have to agree directly with this chart. The Theosophist Scale breaks the physical plane down between 0–1,000, with the lower astral terrain from 0–160, the mid-astral terrain 160–275, higher astral plane from 275–475, the mental plane from 475–600, the Buddhic plane from 600–740, and the atomic plane from 740–1,000. Hinduism has a scale from 200 to 300 to 500, and Lester Levenson has a scale from 50–475.[34] The chart

---

34 Frederick Dodson, *Levels of Energy* (CreateSpace Independent Publishing, 2010). 61

itself is not as important as grounding in the concept that there are different energies in every single thing that we feel, do, and experience.

Whatever scale you use, the important lesson to be learned is that we're able to maneuver through realities and manifest them better at higher levels of energy. Looking back on memorable moments in your life, where powerful or profound things happened, you can see their connection to higher energies. Events like Christmas, a wedding, or the birth of a child happen at a certain level.

At a place of unconditional love, where we allow our hearts to become loving and forgiving and to overcome despair, miracles can happen. At the lowest levels, we can become locked in, where it's difficult for anything else to resonate with us. When we dip down that low, the despair can turn into a complete disrespect for yourself and any other life. Anything below guilt, shame, and despair would be pure evil. On the other hand, when you're higher on the scale it becomes more difficult to move down or to engage in low-frequency emotions.

Notice as well that anger is a higher energy than guilt or shame, love is different than acceptance, and willingness is different than pride. Each of these energies is completely different, like spices in the kitchen. Knowing how they are different and how they function differently is a way to understand how we can better affect reality. Be aware of your subconscious emotions and way of thinking, because the way you view yourself and the world will project an energy that draws in like energy.

At low levels of grief or worry, it's difficult to see the beauty, creativity, and possibility around you. At a level 500 in power you might not be able to experience love unless you move beyond it. You might not be able to find true ecstasy or exultation until you go beyond unconditional love. The greatest awareness happens at the highest levels, when we begin to understand that we are all connected subconsciously to everyone else.

When you meet someone, no matter how you feel about them, some little piece of them is you. If you're angry, the person you're angry about is you. With this perspective, you can look at other people and experience some of their own experience—you can be one with them at the grocery counter or in traffic. Imagine yourself being on the other side, standing behind the register. Imagine being them, remembering that moment. When you reach that level of oneness, nonduality, and awareness—which doesn't happen all the time—it opens access to all possibility.

THE MOST POWERFUL ENERGY STATES

A level of energy that is really powerful and accessible is that of humor and laughter. We can diffuse dark, depressing situations with a little humor to lighten the energy in the room. Books about ghosts will tell you that you can protect yourself from ghosts by laughing. It literally raises the vibrations of energy in a room and can instantaneously change your own energy. For that reason, I believe humor is the most powerful energy that we can focus on and become part of.[35]

Another powerful energy is that of sexuality. This is how we create entire living creatures, and its energy carries that power as well. The explosion of energy in the Big Bang almost becomes a metaphor connected with the explosion of ecstasy and happiness that comes with the creation of each person on this planet. By yourself or with someone else, sexuality is the energy of creation. Napoleon Hill, back in the 1920s, talked about sexual energy as a powerful component of creating our own thoughts.

You might be uncomfortable talking about sexual energy, but we need to acknowledge it as something separate from the other energies and as one you can utilize to immediately change your energy and better maneuver through reality. It is intense and emotional, which we know is more impactful for the way our brains work. It is creative, and it forms links and strengthens connections with other people. Even if you aren't having sex with someone, sexual energy itself has all of these effects.

Back in the 1800s, Aleister Crowley was put in jail for saying sex magic was the secret and for hosting orgies to accomplish that magic. That's not necessarily the take-

---

[35] Laughter is what I consider to be a "mind hack," and it's among many that I've included at the back of the book.

away for use today, but we do know that people have identified the power in sexual energy for generations. There is some kind of "magic" in it, which makes it a powerful energy to tap into.

ENERGY CLEANSING RITUALS

On a psychic and energetic level, energy can and should be cleansed from yourself and your space. It can be part of a meditative practice or a physical action. In fact, just cleaning your house can affect energy. We can see this in the popularity of Marie Kondo. I believe she's doing more than decluttering—she's guiding people toward eliminating negative links to objects in their home.

There are many experts in energy cleansing, and while I won't pretend to be one, I can tell you about techniques that have worked for me. For example, when you get a new house or want to cleanse an environment, go into each corner and clap or make a loud noise. The idea is that the loud noises move the particles and molecules around in that space. After you're done, be sure to wash your hands.

People use crystals, incense, candles, smudging, and smoke as part of cleansing rituals, many of them dating back to ancient practices. Friends of mine will mix alcohol, water, and salt then light it in a room and let it burn until it goes away. In 1930, an occult writer named Dion Fortune wrote a book called *Psychic Self-Defense*, talked about what we can do when we're stuck in a place that we don't have the time or ability to clean. In those cases, you can take your hands and create a circle of intention around yourself, then ask God or whatever part of the universe that you speak with to cleanse this area around you.

A lot of the teachings around cleaning an energy space are just about visualizing the space as clean. Whatever gives you a deep sense that you have cleaned a space probably works. Go with what you feel is best, because then it will be.

Numerous magical and cultural rituals designed to cleanse energy have been created all over the world. Pay attention to these rituals, there may be truth in all of them. When you meditate in any form, your energy becomes higher and purer. Visualizing clean energy around you, through a "magic" ritual or by simply meditating with intention, likely creates a similar result.

Our rituals might help us access vibrations of energy that we have no real concept of yet. We may be able to call upon angels. If our mind creates reality, it may be able to create angels. If people in the past believed in them, why can't they exist now? Invoking energy tied to ancient practices may be invoking the same energy that healed and protected people throughout the generations before us. We can at least be aware of them and their potential effectiveness, rather than simply writing them off because of our own lack of belief.

ENERGY MERIDIANS IN THE BODY

When people talk about energy, chakras often come into the conversation. I've mentioned them and plenty of other people have written volumes on them. Still, chakras unfortunately have a negative connotation for some, and there is a lot of misunderstanding about how they work, what they do, and where they are. Dr. Joe Dispenza refers to chakras as energy centers. I like to refer to them as meridians. In acupuncture and Chinese Medicine, a meridian is a set of pathways in the body along which vital energy is said to flow.

Some meridians receive more attention than others. For example, there is a meridian outside of the physical body, six inches above the crown of your head. Perhaps this is where we access higher dimensions and come into connection with the source. Before any of that, however, we need to open up the energy flow throughout our bodies. This is the intent of focus on energy meridians.

Just below that is the crown meridian, in the middle of the top of the head. People who channel spirits claim that the connection happens through the crown. This is where we find devotion, unity, and spirituality, which is why the crown is often associated with the spirit.

The third eye meridian is, as we've discussed, in the pineal gland. This is more than a meridian. My own experiential theory is that the pineal gland is a gateway between realities. It's located roughly between the eyes and just slightly higher, and it is associated with intuition, the future, and spiritual insights. This is where I believe we access source in the quantum field. Scientifically, it is a ganglion nerve complex that can receive light, activating when we go to sleep and move into theta brainwave states, secreting chemicals and hormones that keep us asleep.

The throat meridian is located in the thyroid gland in the lower neck. When people say they have awakened this meridian it indicates their ability to express themselves, their ability to write or sing, and the part of themselves that is not afraid of what other people will think of their expression. Someone who is very shy might have a constricted throat meridian. A little lower, between the throat and heart meridians, is the thymus meridian, which works with all others. It's associated with compassion for self, forgiveness, and increasing and maintaining vitality and wellbeing.

The heart meridian is an entirely second brain, arguably smarter than the brain in our head. It is not the actual heart, though it is in the center of the chest and near the heart. A fully open heart meridian allows you to fully love. It is connected with compassion for others, acceptance, and forgiveness. From a quantum physics perspective, the heart is the wave energy that particles shift into when they aren't observed. It is all possibility, while the energy in our lower chakras are closer to particle energy.

Between the solar plexus and the navel there is the solar plexus meridian, often called the powerhouse. Charles Haanel, in *The Master Key System,* says that this is where all our manifestations happen. It's the center of personality and gives off a great deal of energy. The solar plexus meridian is associated with personal power, magnetism, and force of will.

Finally, there are the sacral and base chakras. The sacral chakra is between the navel and perineum, and it is associated with emotions and sexuality. It is all about passion, pleasure, and experience. Between the anus and genitals, at the perineum, the base chakra is associated with grounding and how you believe life is taking care of you. It's primal, connected with fear and unbalanced or blocked energy.

For many people, their energy stops here, or maybe just above in one of the lower chakras. They are concerned with survival and fear, which we often feel in the gut—in the lower areas of the body. The meridians can become constricted as well. As you meditate, you can focus on each meridian and how it interacts with the others to allow energy to flow all the way up to the heart, where possibility lies, and beyond into the source at the third eye and above your head where your energy can access infinity. In yoga, inversion poses like headstands and shoulder stands are meant to reverse the flow of energy. You can also use inversion tables or gravity boots to hold you upside down for short periods of time. Not only do these movements lessen gravity compression, but they can be another way to activate energy.

LESSER KNOWN MERIDIANS

There is another meridian as well, six feet beneath your feet. This is the earth star meridian, which people talk about when they suggest standing in the grass at least once every day. This is a practice verified by science—there are absolutely positive effects to putting your feet in the grass. The chakra concept says that we are rooting ourselves into the ground and connecting with the earth.

Finally, there is one more meridian that isn't discussed as much, called the Bindu in yoga and the Plait in transurfing, which is where we become conscious and connect with the Divine. I believe this is where forms of manifestation move from observation and into reality.

Vadim Zeland's *Tufti the Priestess*, introduces the concept of an outer intention plait that starts from the back of the shoulders and branches out like a phantom limb. Carlos Castaneda talks about an "assemblage point" in a similar way. In yoga it is referred to as the *bindu* chakra.

PLAIT

Zeland proposes the idea that something in the universe has recorded all of our possible realities that exist. You can think of them as being filmed while we were focused on the present reality, and that if we shift our focus to one of these other films, we can move into that reality. Zeland proposes that by doing exercises to awaken this energy cord and visualizing and moving our visualization or feeling onto the tip of this energy cord, we are effectively uploading a new movie with a specific reality into our existence.

The plait and the bindu chakra are where we focus to do that.

This region of our awareness is important. As we understand the observer effect, this is the realm closest to our own energy field that is in a wavelike state. Energy may be streaming from this area behind us into our awareness because there is an observer gap or blind spot behind our head. By understanding this energy flow and activating and directing visualizations toward the plait, my visualizations have begun to appear in manifest reality on a regular basis. This is where consciousness is plugged into the space of variations in all realities. Once you focus on the area, you will start to feel sensations from this cord of energy. You then visualize into the tip of the plait as if you are uploading a new film.

What's interesting is that this is not unlike the cords and links of energy we talked about at the beginning of this chapter. As you meditate, you imagine where that plait or chakra might be, and as you connect with it you begin to feel it. If there is something you want to experience, you put that powerful thought or emotion into that plait and embrace it. See it, hear it, feel it, and place that into the energy stream there at the bindu, which maneuvers you into that intention.

GEOMAGNETIC INFLUENCES

Your intentions and manifesting are directly affected by

the energies of the universe; particularly by geomagnetic influences, like the sun and planets. Lynne McTaggart, in *The Intention Experiment*, documents the succession of Persinger studies in the 1970s, which ultimately found geomagnetic activity connected with increased intentions. For example, if someone "sends thoughts" to someone else, the thoughts are sent more readily when there are active solar storms.

Persinger worked with paraspsychologist Stanley Krippner, who had perfected an experiment to test telepathy, clairvoyance, and precognition during deep sleep. In these dream laboratories, for example, they might partner people off and attempt to transmit images to people who were sleeping and dreaming. When they looked back over the data, the dreamers had significantly higher accuracy in picking up the target pictures in their dream recall sequences on nights when the earth's geomagnetic activity was relatively quiet. On days of geomagnetic calm, spontaneous instances of telepathy or clairvoyance were also more likely to occur, and remote viewing accuracy was more likely to improve as well.

It appears that we're able to physically affect objects when there is more geomagnetic activity, but we're able to communicate on a telepathic or thought level better when there is less activity.

McTaggart did note that it would have been tempting for Persinger's test to conclude that all spiritual experience as geomagnetically induced hallucination—except for one unsettling fact: even minor environmental changes, from slight variations in the weather to solar patterns, appeared to have a profound effect on extrasensory perceptions or the ability to view things remotely.

Even if you doubt things like astrology, it's clear that there may be some kind of influence coming from geomagnetic activities—which might be correlated with full moons and star alignments. This also indicates that an attempt at manifesting something through meditation might be affected by geomagnetic activity. If it tells us nothing else, it is a reason that you want to keep trying. Whatever you're doing, do it far more than just one time. We never know what variables are at play.

BUILD UP YOUR ENERGY

During a recent interview on my podcast, physicist and author Cynthia Sue Larson explained that energy is the key to making these jumps into parallel realities. An electron requires a burst of energy to complete a quantum jump.[36]

It's important to understand that we're not talking about the type of energy burst that comes from a cup of coffee. Instead, we have to amplify our *qi* (or life force) energy. I have experimented with several different

---

[36] Brian Scott, "Interview with Cynthia Sue Larson - Physicist, Life Coach and Writer." Podcast Audio. *The Reality Revolution Podcast*. August 23, 2019. http://www.therealityrevolution.com/interview-with-cynthia-sue-larson-physicist-life-coach-and-writer-ep-98.

options to create such energy bursts, and landed on the ones that follow.

THE FIVE TIBETAN RITES

Five Rites routine (also known as the Five Tibetan Rites or Five Tibetans) is used to build up the energy necessary to perform a quantum jump into a new reality. This is a very popular routine of exercises. I was first exposed to it by the wonderful science fiction writer Steven Barnes. Vadim Zeland also does this exercise and gives advice related to its application in his book *Apocryphal Transurfing*.

The Five Rites is said to be at least 2,500 years old, though Peter Kelder popularized it in 1939 with his book *The Eye of Revelation*. His story is that a guy was desperate to find the fountain of youth, and he's told that old men traveled to this place and came back inexplicably strong, healthy, and full of vigor. After retiring, he went to live with those specific Lamas, who taught him these five exercises that they called the Rites.

Rite 1. If you've ever been to a Grateful Dead show, you've seen people spinning around in circles—the Whirling Dervishes. For many people, that kind of an experience is just as spiritual as a church service. This kind of spinning movement is the first of the Five Rites.

Stand erect, with arms outstretched horizontally with the shoulders, and spin around until you become slightly dizzy. Count your spins going to your right (in the

Northern Hemisphere, to the left in the Southern). In the beginning, you might be dizzy, and it's not recommended to do it very much. I have my own theories about how it works, such as mixing up reality in that moment—spinning it all around. In *The Eye of Revelation*, it is described as opening up psychic centers in the body, which are not unlike chakra centers.

When I first started doing this, it felt weird. Then, around day sixty, I noticed that it was generating energy. The spinning motion seems to pull in energy and speed it up, amplifying it in your body.

Rite 2. Next, lay down, full length, on a rug or a bed. Place your hands palm down, alongside the hips. Keep your fingers close together, with the fingertips of each hand turned slightly toward one another. Raise your feet like a leg lift until the legs are straight up—if possible, your feet can go all the way back over your body, just don't let your knees bend. Hold this position for a moment or two, then lower your feet back to the ground or the bed. Repeat up to twenty-one times. At first, you might only be able to do one or two or three. That's fine. There is no need to do more than twenty-one.

After doing this movement regularly, I could feel the energy I had generated from spinning being pulled up to the meridian in my gut.

Rite 3. Kneel on a rug or a mat, with hands out at your sides, palms flat against the sides of your legs. Then lean forward as far as possible, bending at the waist with your head forward and chin on your chest. Next, lean backward, stretching your body and causing your hips to move forward and backward, hands still against the sides of your legs. Your toes will keep you from falling over. Then move into that erect kneeling position, as relaxed as possible, and begin again. It stretches your spine, pulls your shoulders back, and moves the energy backwards and up. See if you notice the energy moving to your solar plexus. Work up to repeating this movement twenty-one times.

Rite 4. Sit erect on a rug or carpet, legs flat and feet outstretched in front of you with the backs of your knees on the rug. Place your hands flat on the rug, fingers together and pointing outward slightly, with your chin on your chest and head forward. Now, gently raise your body from that position with your hands, bending your knees so that from the knees down your legs are practically vertical, matching your vertical arms. As you lift up, let your head gently fall backward as well, as far as possible. Hold, and repeat twenty-one times.

Don't worry. This is admittedly one of the most difficult of the exercises, and you don't have to do it exactly as it's explained here. For example, I have a bad shoulder so I rest the top of my back on a bench and thrust my hips upward. This is enough to move the energy into my heart. Alternatively, put the upper part of your back on your bed with your feet on the ground, then lift your waist up and let it fall back down. The goal is to stretch that part of your body. Repeat twenty-one times.

Rite 5. Place your hands on the floor about two feet apart with legs bent, then stretch legs out to the rear with feet also two feet apart. Push your hips up as far as possible, like the downward dog position in yoga. Drop your chin to your chest and rest on your hands and feet, then allow the body to come slowly down to a sagging position while bringing the head up and back as far as possible. In yoga terms, this movement is basically a downward dog moving into cobra pose in yoga. It's more grueling than the others, but if you can work up to twenty-one of them, the energy really starts to flow.

Once I finish these five movements, I use an inversion table to roll upside down twenty-one times. This movement flows energy and blood toward the head. This movement is optional and may be difficult, but if you can do it, you will notice an even greater energy flow.

Altogether, this routine should take about five minutes total.

FIVE TIBETANS

KNOCKING ON THE DOOR OF LIFE

After the Five Rites, the next thing that I do comes from Qigong. It's a simple step, and the goal is to cultivate high-quality energy. In this, you will start by standing with your feet shoulder width apart, then begin a slow turning motion from the hips and waist. Keep your shoulders, arms, and upper back relaxed—let the momentum from the center of your body move your arms, kind of flopping around.

As you continue, let the momentum continue to pick up your arms until they knock gently across your lower back. This stimulates the pressure point on the spine directly behind the navel, called the Door of Life, or the Ming Men.

This pressure point enhances your overall vitality and energizes the central nervous system. It's a simple exercise that you can repeat, turning back and forth twenty-one times, or whatever feels natural.

Breathe throughout this entire exercise, and afterward, do a qi massage on your lower back. To do this, put your hands on your lower back and massage it with loose fists tapping your lower back. Start as high as you comfortably can, and go as low as you comfortably can. This wakes your adrenal glands and stimulates the Door of Life pressure point. It also energizes the kidneys and brings circulation to the lower back.

Now, move down to the outside of your legs and slap them with the palms of your hands, all the way down to your feet, then back up the inside of your legs. Come up to the hips, still gently slapping your legs, up to a Tarzanesque pounding on your chest. This stimulates your lungs, heart, and thymus gland. It's rejuvenating, to the extent that it is said to slow down aging.

Do this entire circuit three times, and at the end of it shake your body out. You might have seen athletes do this before a basketball game. The players are nervous and have a lot of excess energy, and they shake their body out. You're going to do the same thing, but to shake the negative energy out.

DOOR OF LIFE (PART 1)

DOOR OF LIFE (PART 2)

COMPLETING THE ROUTINE

Another exercise I like to incorporate is called the Tiger, or simply Bending and Lifting. It's another simple set of movements, beginning by bending down as low as you comfortably can. Act like you're grabbing energy from the earth, pull it up to your heart, then raise your hands into the air and stretch up as you pull the energy up and pull it back down.

Repeat this twenty-one times as well, which will get your heart moving as you bend up and down over and over.

You can also try the Zen Swing. To perform this movement, stand relaxed with your knees slightly bent. Draw your arms back, like you're hitting a golf ball. When you come to the end of the swing, allow 90 percent of your weight to shift to your front leg. Allow your arms to swing naturally and you will find that by changing the angle of your swing, you can access any area of your spinal column that needs a little grease or loosening. Continue this relaxing, unwinding swing pattern for as long as you feel comfortable doing so.

ZEN SWING

Finally, there is the Three Thumps from energy psychology. It takes about a minute, which should round the entire process out to ten minutes. Take your fingers and thumb to your cheeks about twenty-one times, then thump down on your chest right above your collarbone, where your thymus gland is. Thump there, then thump at the sides of your body, on your spleen.

Each of these areas has benefits. They are said to lift your energy level, balance blood sugar, increase metabolism, increase energy, increase strength and vitality, and increase alertness and energy.

After each of these steps, sit for a moment and feel the energy and tingling in your hands and your body.

If you're in a hurry, you can skip the Five Tibetans or the knocking steps—but if you can do all three in combination, it's like drinking a pot of good coffee without any of the side effects. The energy in your body builds up, and when you begin to meditate, the electromagnetic energy that you can transmit is that much larger. I believe it's exponential. It magnifies my meditations, my intentions, and literally everything that I do.[37]

You can follow any other energy-building exercise and it will probably work just as well.[38] Go with your heart and your feelings—go with what will work for you. It could be push-ups or sit-ups. It could be running, or anything that raises your energy levels. I follow this routine because it tries to activate different points in the body and stretches you out. It helps you to have good posture when you meditate. The energy is flowing, you're not as uncomfortable, and you're able to sit still and begin the meditation from a good place.

If you want to maneuver realities or even quantum jump into a reality that is much better than the one that

---

[37] Quantum jumps, which are described in the resources section of the book, require this kind of energy to work.

[38] Biohacking and the exploration of food and exercise to build energy, found in the back of the book, are interesting techniques to explore.

you have, it will require energy. Be aware of the energy in your life, in your food, and your exercises—in the links of energy around you and the levels of energy in your universe. If you were to only focus on energy building and flow in your life and surroundings, your life would change. But keep reading, because we have so much more to do.

CHAPTER SIX

TUNE IN TO YOUR INTUITION

"*Technology is not going to save us. Our computers, our tools, our machines are not enough. We have to rely on our intuition, our true being.*"

—JOSEPH CAMPBELL

Back in World War II, air raids were such a regular part of life in London that many of its residents became indifferent to the danger they indicated. They heard the sirens so often that they started to act casual about them. In the book *Mysteries of the Unexplained*, Prime Minister Winston Churchill was described as "the embodiment of Britain's unyielding resistance to the enemy," and notes that he took his role seriously. He was just as indifferent to the sirens as anyone else, perhaps more, because of how disrupting the sirens could be to his work.

Yet the book goes on to say that Churchill listened closely to his inner voice. One night, when entertaining government ministers at 10 Downing Street, the traditional residence for Prime Ministers, an air raid howled throughout the dinner party. Then, right in the middle of the meal, Churchill got up and went to the kitchen, which had a large, plate-glass window. He told the butler to put the food on a hot plate in the dining room, ordered the kitchen staff to go into the bomb shelter, then returned to his guests.

Three minutes later, a bomb fell behind the house and completely destroyed the kitchen—with the Prime Minister and his guests miraculously unharmed.

On another occasion, when Churchill had just finished visiting an anti-aircraft battery during a night of attacks to inspire and buoy the fighters with confidence, he walked over to the car that had his side door open and waiting for him. Instead of getting in the car, he walked around to the other side of the car, opened the door and got in there instead. As they drove off through the blacked-out streets, a bomb exploded that lifted the car and destroyed the side that he was meant to be on.

When his wife asked him later about what made him choose the other side of the car, he first said he didn't know, then corrected:

> Well, of course I knew. Something said *stop*. Before I reached the car, the door held open for me, it appeared to

me that I was told I was meant to open on the other side, to get in and sit there. And that's what I did.

When we talk about intuition, we have to realize that history has been utterly guided and created by incredible people who trusted their intuitions. Throughout history, great men and women are great because they have trusted that voice. Everyday people have had their lives spared because they didn't get on the train that crashed, walk into the building that was bombed, or get in the plane that would go down.

According to a study by W.E. Cox back in the 1950s, fewer people travel on trains that are destined to have an accident than on trains that are not. These accidents are completely unpredictable, unknown until they occur. He found that many people consciously or unconsciously avoid taking a train on the day that it is going to crash. How is this possible?

A lot of people want to factor magic into this, but there might be something more. Another study, this one by Gerard Dietrich Wasserman, a mathematical physicist at the University of Germany, England, says that "all events exist as timeless mental patterns, with which every living and nonliving particle of the universe is associated. We are aware of all these events at the same time." In other words, there is part of us that knows everything that is and is to come—it is the intuition, and it's something we can all tune in to.

WHAT IS INTUITION?

In preparing for this book, I asked some of my friends how they experience intuition, and their responses varied. Some of them got a flash in their minds—have you ever perceived a meeting with someone by their photo or because their name rings a bell, but you've never actually met them? Some got a feeling in their stomachs—butterflies of recognition about a place or a person that you've never actually seen before. The majority of them said it happened in a quick moment, and many of them didn't trust their intuition at all.

In these flashes of knowledge, our bodies might be signaling safety or danger from certain places or people, pulling from a pattern of knowledge that we don't consciously understand. Maybe it's remembering something from the past—maybe it's remembering something from the future. Mostly, we tend to ignore that activated knowledge just so we can stay sane. Someone told me I should invest in Google back when it was five dollars per share, and my intuition agreed. But I didn't do it. You might be able to relate in some way—perhaps none of us really follow our intuition as closely as we could. If we understood intuition better, we might be able to follow it better as well.

So what is intuition?

The dictionary gives intuition synonyms such as *hunch*, *premonition*, and *clairvoyance*, but we don't have to be psychic to have intuition. Sometimes it's easy to take those synonyms and conflate intuition with superstition—don't leave your hat on the bed, or watch out for black cats walking in front of you. While I suppose those could be connected with intuitive signals that people had in other existences, that's not the kind of information we're trying to tune in to. We're looking for the little hunches that tell us not to get in on that side of the car to guide us into the future.

Adrian Dobbs, a mathematician and physicist at the University of Cambridge, looked at the nonlinear "lake" view of time and proposed that there are actually a relatively small number of possibilities for change that exist at a subatomic level, and that especially sensitive people can sense when a wave has disturbed those possibilities. They can sense the toy ship coming to the edge of the lake, to call back to our discussion in chapter 2. They can use their intuition as technology, as Gavin de Becker says, to measure and identify what might be coming. The catch is that the technology usually operates on a subconscious level. It happens instantly and without conscious reasoning.

John Kounios, a psychology professor at Drexel University calls it "the product of unconscious information processing." Sophy Burnham, author of *Art of Intuition*, describes it as a "subtle knowing without ever having any idea you know it." Consistently, definitions of intuition point to knowledge that doesn't come from the rational mind.

So can intuition, an irrational concept, be scientific

at all? Hundreds of experiments have asked that same question, and have come to measurable and identifiable conclusions. Intuition is more than just a little hunch. It's something that can guide us to an incredible place in the future. It's how we know to get in on the other side of the car. It's how we survive. It's how we live better lives.

Intuition is just information coming in, filtered by our brains and making it through when there's a message or warning we need to hear. It comes to us precisely when we need it, not a moment too soon or too late.

What we're talking about in this chapter is not "woo woo." It's not crazy, and it's not magic. Instead, as Joseph Campbell said, it's what will save us.

THE SCIENCE OF INTUITION

The gut feeling that many dismiss as coincidence is, in fact, a cognitive process moving much faster than we can follow. Intuition is far different than the familiar step-by-step conscious thinking that we rely on so willingly that we think it is a better thought process. In fact, intuition is like soaring flight compared with plodding logic. And, as part of nature's greatest accomplishment, the human brain and its incredible intuition is never more efficient than when its host is at risk.

In *The Gift of Fear*, Gavin de Becker said that intuition "is the journey from A to Z without stopping at any other letter along the way." In moments of fear and danger, intuition is catapulted to another level entirely. Research around pattern recognition and decision-making paints intuition as a sixth sense of sorts, in which humans can act in unique patterns without consciously, intentionally analyzing them, and even more so in situations of fear or anxiety.

The Columbia University Medical Center found that participants' brains displayed anxiety when they saw faces that expressed fear, *even when* the images flashed too quickly for the viewer to consciously see them. In another experiment known as the Iowa Gambling Test, participants had negative emotional responses to a losing card bet long before they became consciously aware that it was a bad bet. Other studies demonstrate test subjects performing tasks better when their conscious minds were distracted than when they paid full attention.

Clearly, the subconscious is intricately involved with intuition. The unconscious mind is like a quantum computer going through millions of bits of information, processing what we cannot, then often sorting through that information to give us insights while we sleep.

Rebecca Spenser, assistant professor of psychology at the University of Massachusetts, found that REM sleep helps subjects with problem-solving and decision-making, and the more REM sleep you get, the better it works. Why? Because our unconscious processor—intuition—works better.

Sometimes, intuition is better than logic, as demonstrated by a study that told women to "eat intuitively." They relied on an internal, subconscious meter to determine whether to eat and how much. By the end, those women were found to have lower BMI than women who went on calorie-restricting diets. In a similar study, the *Journal of the American Dietetic Association* found that women who practiced intuitive eating over two years maintained their weight and had higher self-esteem and higher levels of physical activity.

In business, innovators like Steve Jobs, or Tumbler and TweetDeck investor John Butterworth followed their intuition. They followed their intuition to such great success, that there are now intuitive consultants whom you can hire to help you follow in their steps by following your own intuitive path. Sonia Choquette, author of *Trust Your Vibes at Work*, suggests you gather all of your facts logically, then turn them over to your intuition to percolate.

We use intuition interpersonally, as well. *Psychology Today* calls this *empathetic accuracy*. Our ability to read others goes beyond interpreting words and emotions. Researchers Radin and Schlitz, who wrote *Gut Feelings*, did a meta-analysis of previous studies and concluded that people mentally influence each other's psychological states that transmit conventional models of human interaction. Simplified: people influenced the gut feelings of their partners, who were in separate rooms, fifteen meters away.

You might be thinking of the energy connections that

we discussed in the last chapter, and you would be right. Intuition works at that quantum level, and certainly will be influenced by energy and those links we have to other people and things.

When intuition tells you something, it might be based on valid, absolute, real evidence—but there's no guarantee that you'll be able to verify it. Rarely will we have time to go through all of the information in that moment. One exercise that Gavin de Becker walks through is called "two minutes," where he analyzes two minutes of a given, historical assassination attempt, second by second. When those tragic events happen, we only get those minutes once. We can't consciously pore over millions of bits of data in that minute—but the human brain can access it in seconds or less to trigger a life-saving or life-altering reaction, like a measurable miracle.

INTUITION AND PREMONITIONS

Imagine going into a place in your meditation where you have the memory of everything that ever happened, in the past or the future, and the thoughts of everything along the way. You aren't quite able to process all of it, but it's there. You exist in it. When you come back to your normal life after the meditation, you already know what's about to happen. You've already experienced it through the eyes of everything and every living creature around you, all at once. You have accessed the source field that carries infinite intelligence.

We have talked about the source and what it might mean to have access to that information, but it's more than just a theory. Larry Dossey, MD, said in *The Power of Premonitions*, "Premonitions suggest we are linked with every consciousness that has existed, does exist, or will exist. That we are part of something larger than the individual self." Physicist David Bohm explained that "Each person unfolds something of the spirit of the other and his consciousness."

We are connected not only to each other in the present, but to all time and reality. Once we connect to this oneness in our meditation, we begin to access intuition in a stronger way. We have already experienced these things in the future and the past. So while it's part of our survival skills, intuition doesn't have to just protect us from dying. Our soul has access to the whole information field and can see what lies ahead, even into future sectors that are not realized yet. The soul knows what is good and bad for us and how to guide us into the incredible futures that we have set for ourselves.

Irving Schwaninger said, "There's obviously only one alternative—namely, the unification of minds or consciousness. In truth, there's only one mind." To ignore intuition as a way of knowing would be like driving a car while wearing blinders. It is critical.

As humans evolve into their next higher selves,

as we experience this reality revolution that we're all going through, we will gain more intuition, not less. We will carry our old mind/body reactions with us as we progress from animal consciousness to spiritual consciousness. As Steve Jobs said, "Have the courage to follow your heart and intuition. They somehow already know what you truly want to become. Everything else is secondary."

THE EXCHANGE OF ENERGY

Our eyes are programmed to see objects as three-dimensional and solid, and our nerve endings are programmed to feel them as three-dimensional and solid. But in reality—in the quantum domain—there is not solidity. When two clouds meet, is there solidity? In the same way, everything is clouds of particles that are melded into quantum coherence. Everything melds and separates and melds again into something similar.

That means, whenever you touch an object, you give it some energy and you gain some of its energy. We're constantly gaining information from these energy fields. The electron clouds meet, small portions meld, and then you separate. You still perceive yourself as whole, but you have lost some energy or gained it.

Intuition, on the other hand, is also receiving information from an information field—a template for what things could be like. When energies within that template have the same resonance, that part is melded and transformed into physical reality.

The conscious mind might not be able to read that information, but the subconscious mind accesses it directly—which is where those hunches, premonitions, prophecies, inventions, and works of art originate. *Intuition.*

THE ACCURACY OF A HUNCH

I owe much of my success and happiness to intuition. Many of my decisions are made in opposition to normal reasoning. Intuition does not explain or justify its reasoning, however; it only points the way. If you set out to find an explanation, you'll start to lose your ability to tap in to intuition in the first place.

Let me give you some examples. I get an email from someone who asks me to do business, and I intuitively know whether or not I want to deal with them—even before finding out more about them, checking Google, or having a face-to-face meeting. I instantly know. Or, new software is in front of me, and I can intuitively click around to learn how it works before reading the manual. In the instant that someone offers me an investment in the stock market, I can tell whether it will work or not. The instant someone hands me a book, I can tell if it's good or not.

Once, I left to pick up some things for my girlfriend, and I found myself in a rush once I got to the convenience store. After I left, I saw the very store I was rushing through on the news. There had been a hold-up at the store, and it happened at the exact time that I left. Looking back, I can remember that moment, the burglar walking in as I walked out. I remember he wore a jacket even though it was summer, and there were people sitting in the car that he came from. Perhaps my intuition could read the situation, while I couldn't pick up on those things until later.

I've had multiple examples where trusting that first flash of insight has saved me, not the least of which being the home invasion story where I knew to shut the door and run. But there's a dilemma around that example: Why didn't my intuition tell me this was going to happen in the first place? Why didn't I lock my doors? Why didn't I avoid it in the first place?

These are questions I've had to grapple with, because I have tuned my intuition. I believed I was starting to tap in to its power. We'll look into that more in a moment, but for now I acknowledge that this reads as though it is a power. I'm aware that some people will consider this paranormal, but I believe that separates us from the truth of intuition. It is simply attention that can be expanded to places outside of the visible environment. The more you expand your awareness, the better you become at knowing what's going on.

INTUITION AND COINCIDENCE

Have you ever seen someone, maybe on your Facebook feed, say something like, "It's 11:11! The angels are communicating with me!" Or tell you that they thought about a woman in a black dress then saw a girl in a black dress and *must have manifested it!* Let me just say this: no, you didn't manifest anything. While I have had enough 11:11 occurrences that I enjoy looking for them and I sincerely hope that angels are guiding me when I see them, I assume there is always a chance that my reticular activating system (RAS) is at work. As humans, we are really good at finding things when we start looking for them.

What we need to understand in those cases is the difference between coincidence and synchronicities. That's where the reticular activating system of the brain comes in. Remember, our brains are incredibly complex and can sift through billions of bits of data at any given time. How do we survive in that environment without short-circuiting? We organize the data.

The reticular activating system is the main filter through which all of this information enters our bodies and minds, through a bundle of nerves in the brain stem. The important stuff gets through, including data from sensory organs and anything that you've signaled is important. You're telling the RAS what to look for. That's why it alerts you to 11:11 or a type of dress you were thinking about.

As you become aware of this power, you can start to

use it with intention. If you're a writer, you might focus on a particular topic then notice a flood of information on that topic coming from everywhere. That's because you're paying attention to it, so the RAS starts flagging it for you. The truth is, none of it just showed up for you as manifestations. You just didn't notice it before.

The RAS is why you go shopping for a car then suddenly notice it everywhere. It's why you learn a new word then hear it all the time. It's why you can tune out noise in a crowd full of people but hear your own name spoken from across the room. Think about how incredible this is: there is a part of your brain that can hear every single conversation in the room, zoom in on your name, and alert you to it.

When we access intuition through the subconscious, we're accessing all of that information as well. Day to day, however, the RAS creates a filter and only presents the pieces that are important to you. Michael Hutchings talks about this in his book *Megabrain*, noting that once outside stimulus is reduced and we close our eyes, your RAS turns inward. With no other data coming in, it starts to filter out information from your subconscious to give you intuition about what you have experienced.

That's the scientific understanding of the power of meditation. You're simply redirecting your reticular activating system to work for you in another way. This is a wonderful tool that we often relegate to thinking we manifested a woman in a black dress—conflating coincidences with things that have been filtered through our minds.

To train the RAS, we can marry subconscious thoughts with conscious ones. For example, setting an intention means focusing in on a goal. At that point, your RAS will, on a repeated basis, start looking for themes that guide you along that path. We set goals so that we can tune ourselves, through the RAS, to them and move in that direction.

THINKING DIFFERENTLY ABOUT COINCIDENCE

Synchronicity, coincidence, miracles, good luck—these are different terms for the same RAS phenomenon, but it's important to understand that the intelligence of the body works through that synchronicity. When you begin to see coincidence as opportunity for creativity, every coincidence becomes an opportunity to become the person the universe intended you to be.

Predictably, we maintain the same routines day after day, set our minds on a certain course of action, and fall into that rut over and over again. If we march along mindlessly unaware, we don't notice coincidences. How many miracles can really happen if we're tuned out zombies?

A friend of mine is a recovered alcoholic who had drank heavily and got into drugs. I asked him about how intuition played a part in addiction, and he said that a voice told him that it was okay to drink and a voice told him he wanted to drink. He started to figure out how to listen to all the voices. Sometimes the voices deceptively played the part of intuition, but that's not what they were. The mind is not intuition. It is a lonely character with no one to talk to. If the voice is selfish, always saying *I, I, I*—it's probably just the voice of your own brain. When we learn the voices—like RAS, our own tendencies, and intuition—we can spot the coincidences, miracles, and manifestations much better.

Your intuition exists in your heart. It's where you can hear the voice of your true intuition.

LEARNING AND PRACTICING INTUITION

It would be wonderful for our children to be taught about intuition in schools. Instead, we're only finding it as adults, often through meditative practices. People who meditate and are mindful better understand how to tune their bodies and read the messages that their intuition sends. Even Gandhi said that he began to recognize the voice "that prompts us to do a certain thing" when he started to pray regularly.

Not everyone will say that they are intuitive, but many of us can relate to stomach pains when something is wrong, or a sense of heightened awareness that we can't put our finger on. If we were to imagine intuition as a living thing that we couldn't see or hear, but it needed to communicate with us, what might it do? It would look for ways to communicate—say, a stomachache or a sense that something is wrong.

We need to learn to figure out not only the message, but how to trust that message. Sometimes, trust isn't there because we haven't experienced honesty. By being honest with ourselves and with other people, we can begin to trust the information that we have and that we share—and then eventually trust the information our intuition delivers.

Sometimes literal, conscious information can help us. Malcolm Gladwell notes that you become more intuitive when you're in a field that you understand well. On my computer, I'm highly aware of when something isn't working right. My dad was a veterinarian for forty-five years, and he would know right away what was wrong with someone's pet. The blood tests and X-rays would always confirm his intuition, because he recognized the subtle signals that came from thousands of hours of experience.

We don't always have the experience to recognize that information, though. In Australia, about twenty people are killed by kangaroos every year. In all likelihood, if they understood the signals a kangaroo was giving off—for example, bared teeth is a grimace, not a smile, they won't attack with a baby in their pouch, and they'll usually look behind them before they come in for the kill. If you have that information, you'll notice your intuition says it's time to go. If you don't, then the signals might feel illogical. Penny Pierce says, "Information, as well as openings to greater knowledge, can come in strange and illogical ways. We're not as practiced in recognizing and validating intuitive truth as we are a concrete, scientific food."

Maybe you need to take a journey. Maybe you need to do something radical. Sometimes you need to take those chances and follow your intuition to make great things happen.

Remember, intuition is simply comprised of signals from the heart. Once the brain gets involved, that knowledge and truth becomes simply interpretation. All living beings have their own way of interpreting those signals and interpreting the truth. Understand that there are levels of accessing information, then filtering it with minimal distortion. It takes practice, trust, and belief to gather information from the subconscious and maintain it in its purest, objective form. In the rest of this chapter, we will look at ways to practice tuning in to intuition—to open the doors to the information the unconscious mind has already accessed.

WAKE YOURSELF UP TO RESET THE SCRIPT

Intuition comes from outside of the scripts. You might have already noticed moments that were "off script": if you can recall a moment of sudden inspiration to solve a problem, think about where it might have come from and why.

I once helped coach a counseling student who had to make some decisions about where to work, and he started a therapeutic process of examining the imagery in his dreams. After that, when he went into his therapeutic practice, he found that he had profound insights about the people he was working with. He was not only dreaming about work, but his subconscious intuition was processing the actual day to day of his job and helping him in his practice.

Sometimes the information is much more accessible. An art director that I worked with uses the auditory sensation of intuition to sense when a mistake is in the works. He actually hears a little sound that indicates something is wrong with a print run or a project's finish line.

Brian Larson, the creator of Mettaverse Music, sees his music as colors due to synesthesia.[39] By using multiple senses, Brian can see what is musically necessary to complete the feeling of a song. As a result, his music has an eternal visual quality.

Your senses may be mixed in a similar way. Every sensation from smelling to feeling is critically important to understand what is happening in your environment. Enhancing your ability to call upon multiple senses is like learning an advanced mental martial art. It increases your level of awareness and generally maximizes your effectiveness in life.

I'm reminded of Friedrich Nietzsche, who said,

---

39 Link to Mettaverse Music: https://www.youtube.com/channel/UCyvjffON2N0UvX5q_TgvVkw

"Those that dance are seen as insane by those who cannot hear the music." If you hear intuition in a way that other people cannot, it doesn't mean you're insane. Notice what you're feeling and acknowledge it. Wake yourself up from the movie and intuitively choose your own scripts.

INTUITION CHECK[40]

Premonitions: Have you had visions—a dream, fantasy, or meditation—that gave you information about your life, the future of others, or the future of something about the planet? Have you had a déjà vu experience where suddenly you realized you'd already seen everything happening before? Have you ever been awakened by the ringing of a telephone or a knock at your door, only to find that no one was there? Have you ever seen the image of a friend before you unexpectedly ran into them?

Problem-solving: Have you ever caught yourself humming a tune and realized the lyrics were the answer to a problem that had been plaguing you all day? Do you hear music too high for the human ear?

Vibrations: Have you ever felt cold and clammy when entering a strange house? Can you hear tension in the tone of a person's voice? Do certain people make you feel angry or highly nervous

---

[40] These signs and more are taught in Penny Pierce's book *The Intuitive Way.*

> before they even say a word? Can you tell when your pet is upset? Have you ever felt like someone touched you but when you turned around no one was there? Has your hair stood up on the back of your neck, or have you gotten a "creepy" feeling around certain individuals?

CHECK IN WITH YOURSELF

Several exercises have worked for me and people I've coached, beginning with one that's very simple. It's the intuitive awareness that I learned from Frederick Dodson, and that is to just calm down.

If you have a particular question that you need your intuition for, calm yourself through meditation or some way of relaxing, then write your question down. Once you're relaxed, write down every possible answer you can think of, and imagine each as being true. Check in with how it feels. Cross out each item on the list until you get to the answer that feels most true. It's not accurate 100 percent of the time, but it's a good way to start the thought process and to access feelings in your body.

Once, I went out to my car, and a very expensive, very unique pen was sitting in the passenger seat of my car. It had a British guard on it, and I don't know anyone from Britain. I started to think through the possibilities, like whether someone would have intentionally done that, whether my girlfriend was playing a joke, whatever I could think of—and then I remembered that a new family had moved nearby from Britain.

That option felt the most true, so I went over and asked the father, "Were your kids in my car?" The kids laughed but said no, so I went home. But later on, the dad came over to me and told me that the kids had admitted that it was them—they had climbed in my car, which I have a bad habit of leaving unlocked, and they apologized for it.

I might have only uncovered basic information that I already knew, but the process helped me check in with my intuition to come to a decision. Once you master this skill, you'll have the ability to focus on a target with laser-sharp accuracy while keeping your mind clear enough to absorb the information you're receiving.

FIND A NEUTRAL, PRESENT-TENSE BRAIN

Think about it—if you're skiing down a mountain at 100 mph, you have to rely on intuition. You experience the feeling of movement and the flow state that can get you safely to the bottom. There is no way to rely on a conscious, animalized stream of information in that situation. Standard predictions of the future cannot apply.

What we think we know, what people will say, what our worries tell us—if you're not in a calm state, you can't rely on any of that. The ego part of you cannot discern intuition well because it has been programmed with all sorts of information coming from outside sources. To get in touch with any sort of true intuition, you have to get in touch with the unprogrammed part of yourself.

Both the left and right sides of the brain are important, but they do have different functions. One side is creative and the other is logical. If you're overly emotional or overly logical, you might have a dependence on one side of the brain versus the other. For intuition to work, especially when you're manifesting reality, you need to adopt an attitude of neutrality. Whatever you support or oppose will come to you—observing with neutrality can bring greater clarity and knowledge.

This is difficult to do. Deep feelings about certain subjects can make neutrality hard to access, because our brains would rather move forward. The brain is a prediction machine, like the most complex algorithm, and those predictions have allowed us to survive as a species this long. It's in a perpetual future tense, always trying to look to what's coming next. When something violates our expectations or creates intense novelty—a trip, an intense spiritual experience—our brains get stuck on them and struggle to predict the future. *This is not a bad thing.* The algorithm is prediction, not intuition.

Neutrality, stepping outside of left/right brain dichotomy, and breaking from routine can pull us out of our standard algorithms and open up the field of intuition. When your expectations are violated, your senses are heightened and fully engaged, and intuition is accessible.

Flow states are not in line with prediction of the future, but the opposite—athletes and artists can turn off that part of the brain that lives in the future and settle into an intuitive grasp of the present.

Intuition, then, is the ability to rely on the truth you feel in your body in a relaxed state. Our stressful environment makes it difficult to identify that level of relaxation, but it is our natural and truest state. It is an openness that gently observes and is completely unworried. It is the childlike part of yourself that has never been touched by the ups and downs of life, holds a broader perspective, and can actually be aware of the truth.

INTUITION CHECK

One way to access a relaxed part of your brain that is separate from normal patterns is to ask questions of yourself. Start by asking questions that put you in a more intuitive state of mind, such as, "When was the last time I genuinely smiled?" "When did I last sing, dance, or tell stories?" Those higher spiritual states unlock intuition, which is the better part of us. Asking questions like this helps us act as though we're in an intuitive mode, and then we become intuitive anyway.

Next, try to separate from your normal brain patterns using circles. Write down a question in a positive frame, such as, "How can I be happy?" Next, activate your subconscious by circling the question over and over again while you think about the question.

After some time, answer yourself from a higher perspective. How might your confident self answer the question? As you attempt to answer, lay down and practice circular breathing—breathe in to the top of your head as though you're bringing information from the outside, then breathe out through your nose. Keep breathing in and out without pausing, for a few full minutes as you totally relax and imagine your higher self answering the question. Let go completely. Often, the answer will come instantly and undeniably.

SEPARATE YOUR EMOTIONS

Before I moved back to California, I started to intuit that my girlfriend had been texting with some other guy. I'm not a jealous person by nature, but one weekend, she went to run a marathon, and I got a sudden sense of jealousy. The emotion didn't make sense, so when she got back from the marathon I asked her about it. Of course, she initially denied it, but ultimately it was true. She met her first love from high school, they started texting, then she met up with him and they had kissed.

Maybe it's because I was connected to her on an energetic level, but my intuition told me what was happening. If I were normally a jealous person, that hunch might not have been as accurate. Previous experience or acquired knowledge about yourself or a situation can help you make better judgments.

The catch is that we all have multiple voices—the childlike trickster in us all, our emotions, our worries. How do we distinguish, for example, between paranoia and intuition? Jealousy and intuition? What about wishful thinking? The simple answer is that when you're in a positive and relaxed state, your hunches will mostly be true.

Paranoia is a negative, psychotic energy that comes from constant fear, to the point of obsession. Fear itself, while it's a servant of intuition, can become destructive at the point of obsession or an extended state. Some Darwinians believe that early humans were most likely to survive in a state of hyper anxiety. If I were to drop you off on a deserted island or in an environment you know nothing about, your intuition would start to kick in. But while fear can be a signal of intuition in a brief moment, love will take us to much greater levels.

When you're caught up in worry or strong desire, you're likely experiencing emotion-based projections. If you're simply experiencing a slight concern or suspicion without it dominating your emotional and mental state, it's more likely to be intuition. Without keeping those feelings calm and at bay, especially negative feelings, something that starts out as intuition might bring up wanton paranoia. For example, intuiting that someone is trying to hack into your computer gives you knowledge about something that you can fix. If you fixate on it and cover up your computer in tinfoil because some satellite is beaming into your home to steal all of your information, you've let that thought become its own program.

A lot of times, the information we have built up about someone or something completely paints over the intuited information. This brings us back to the beginning of this exercise: the only way to feel the subtle energy of intuition is to relax. Intuition comes when you look at the situation as though you're seeing it for the first time, devoid of emotional baggage, from a mind state of neutral love.

Having life on this earth is a rare privilege that we should all adore. It's an opportunity to practice choice, love, and to overcome challenges. We are all incredible, with amazing abilities to shape reality around us. We're each here for a reason, with unique purposes, talents, and intentions to give the world. Intuition is here to guide us into that future and that purpose.

You don't have to be meditating. You can be driving or hanging out with friends or watching TV. With the proper amount of discipline and commitment, you can train yourself to step back into the necessary concentration levels at any moment.[41]

---

[41] Sheila Oestrander and Lynn Schroeder wrote an excellent book called *Super Learning* that has several exercises around relaxation.

INTUITION CHECK

Mel Robbins wrote a wonderful book about intuition called *Five Second Rule*—and honestly, it may be faster than five seconds as a rule. The idea of instantaneous intuition is to ask yourself a question about something, then give yourself five seconds to answer. It's a simple idea, but it works.

Now, part of you will come forward—the fearful part of you—and will start to create extraneous information and block you from what is true. If you ask yourself, *Should I go on this adventure? Should I write this book? Should I sing this song?* almost every single time, your initial reaction will be right. The ego and fear that creep in afterward will be wrong.

As you learn to find and trust the voice of your intuition, good things will happen, and you'll be able to trust it even more. Just as we wouldn't let a baby drive a car, it takes time to learn to move and walk and become responsible before we can fully earn and develop our intuition. A good place to start is by asking yourself what your purpose is in this moment. I fully believe that when we focus on purpose instead of fear or survival, we'll be rewarded with clearer intuition.

INTEGRATE YOUR SENSES

According to Deepak Chopra, people who are sensitive to events and stimuli around them will be sensitive to "coincidences" from the universe. The clues won't always come in the mail or flash across the television. They can be as subtle as the smell of pipe smoke wafting through an open window—which makes you think of your father, which reminds you of the book you love, which draws out some important thing in your life in that moment.

You might also notice intuition in the way your body reacts. That's the idea behind muscle testing—a form of intuition check. The body is strong when something feels true and weakens when something is negative, not good, or wrong. What's difficult is distinguishing between the responses that the body might have.

To fine tune that sensitivity, focus for one minute at least once a day, on all your senses. Perhaps as a form of survival, our brain blocks out some of the flood of information that we're constantly receiving. Work on tuning in to some of it. Close your eyes. Notice sight, hearing, taste, touch, and smell. Pay attention to each sense and become aware of the information they are bringing in. Allow yourself to take in as many different aspects of your surroundings as possible.

One of the best exercises to heighten your senses is to break down the combination of senses that tend to limit what we see and hear. So, for at least thirty minutes, spend time with a blindfold on. Some people spend as long as an hour or two in this state. After a little while, you'll start tuning in to sounds you don't normally hear. Then, you can use silencing headphones or earplugs to start noticing things that you don't normally see.

Similarly, we can use sound to affect everything around us. Becoming silent ourselves seems to tune the brain in to the sounds around us more. Or, we can use radio to begin to control reality. When you watch a movie, the soundtrack tells you how the story is playing out and even cues your intuition to the structure of the story. If our lives follow scripts, then it makes sense to create our own soundtracks.

For example, you can spend a week only listening to music that you have preprogrammed. Create a playlist with about twenty songs—ideally they will be scores rather than songs with distracting lyrics—then only listen to them in that order. It's a small way to create reality, three minutes at a time. It keeps us aware of the script and the characters all around us, reminding us to wake up from the movie and start to shape something new.

ANALYZING AND USING INTUITION

Sometimes, it's really clear what intuition is signaling and the results it brings. You might be compelled to play a lottery number that ends up winning, or you might have a flash of a business presentation that does well. When results from intuition inevitably come, analyze the sensations and thoughts that you had. Don't let it disappear. In fact, if you can write down any potential intuitive moment, that feedback will help to hone your response to intuition in the future. Set aside a specific time every day to write about the twinges, hunches, stomach lurch, and coincidences that you experienced. Awakening awareness can set us up for better decisions in the future.

It's okay to realize that you have missed many signs of intuition. I once read about a person who went silent for seventeen years, and in that silence, he found the voice of his intuition. Even then, he said, at least 10 percent of the time he couldn't tell what he was hearing. There have been times where I didn't listen to my intuition and ultimately regretted it. That's okay. Those are also instructive moments. Don't beat yourself up over a job you shouldn't have taken or a friendship you shouldn't have started or a business deal you shouldn't have done. Intuition is difficult to identify and even more difficult to trust.

This is less like a science and more like a game of baseball. Sometimes you're going to swing and miss, and that's okay. The more you practice, the more access you'll get—not only to intuition itself, but to the reality you want to shape.

The longer you live, the more puzzle pieces you'll fill in on the way to your higher self. The difficult experiences, the small choices, the big decisions will all combine to form the bigger picture of your reality. Each moment comes with so many choices, and intuition can connect us with the one that really shines. Get to know that voice pointing you to the reality that feels the best. Move beyond intuition for survival into guidance toward things that would truly fulfill you.

The more we understand the way this information is available and presented to us, the better we can recognize that psi-tronic wavefront and maneuver through them to the realities we want.

CHAPTER SEVEN

DEFINE YOUR DREAM LIFE

"People fail in life, not because they aim too high and miss, but because they aim too low and hit."

—LES BROWN

What do you want from life? There's a misnomer, often coming from every other self-help book talking about dreams and intention, that we each have one purpose in life. That we're on some magical quest and we're meant to wander until that singular purpose is revealed to us.

In Victor Frankl's book *Man's Search for Meaning*, he talked about the search for meaning inside concentration camps. In that place where people knew they could die any day, what does purpose look like? What does life look like? The specific purpose didn't matter, but the people who had one lived. The people who lost meaning died.

He continued to explain that you could tell when someone had lost their purpose. In the camps, cigarettes became a form of currency. They could trade cigarettes for a little extra food or something they needed. When someone was in the corner smoking, you knew they had given up. There was no reason left to trade because there was no reason left to live. Those people wouldn't even die by the gas chambers or execution—sometimes they would simply die.

There was no singular purpose that they needed, because our meaning for life adjusts from day to day. What's important is having a reason to wake up and move forward each day. It can literally mean the difference between a life that's wonderful or terrible—between life and death itself.

What would have to happen in the next year for you to look back and say that it was your most successful and fulfilling year yet? Too many of us don't have an answer. We just want to fit in, sleep in, stay safe, and survive. Yet most heart attacks happen between six and nine o'clock on Monday mornings, when stress, worry, and fear are pulling people out of bed instead of vision and purpose.

I believe that there is a particular future timeline for each of us that lights up especially well with our skills, abilities, and goals. Each of us is like an individual planet with our own gravitational pull. When we align with the destiny we want to orbit in and the moons we want to shine in our nights, our lives become more meaningful.

Finding that timeline can feel magical—things feel easier, simpler, and fun. Imagine jumping out of bed when the alarm goes off because you're excited about what's to come. Imagine having a desire to wake up and do things. Imagine having a real reason for living.

Few of us ever reach that point, where we can live in the future that we want, solve the problems that we're uniquely positioned to address, or to see the solution that no one else can. Why? Because we don't dare to shoot for goals that are that high. We decide that our purpose is small and can't take us any further. We talk ourselves out of our uncommon dreams. We gravitate toward what is normal and easy—what our parents, friends, spouses, and bosses want from us.

I have no doubt that when we choose to follow the masses, stay close to the safety net, we all have the ability to hit our goals. But it's only because our goals are small. Is that really the life you want? Our doubts become our traitors, as Shakespeare said, that make us fear to attempt our destiny. How many times in our lives have we missed out on inconceivable opportunities because we were afraid to attempt to pursue them? Because we settled for normalcy?

Looking for a life's purpose is really looking for a way to spend our time in important ways. Not to give us a destination but to give us a roadmap to follow. It takes us beyond manifesting a car that we want or a house that we think we need. It's the understanding that this is a journey. Our lives become richer and more wonderful when we create a vision in line with what we love, then move toward it while we still can.

The larger than life people in history—Martin Luther King, Jr., Gandhi, Steve Jobs, Abraham Lincoln, Rosa Parks, Michael Jordan, Ralph Waldo Emerson, Peyton Manning—made the most of their time. They didn't do what was normal or easy. They were unusual, each in their own ways, and didn't try to be anything else. They were true to themselves and their wild imaginations. And they altered human history.

These heroes of ours breathed the same air as us. The same sun warmed their faces. Yet we treat them as though they are superheroes. They weren't any odder than the rest of us are in our own ways. They became legendary instead of ordinary because they had a vision. They knew what they wanted and how they wanted to fill the world around them with meaning and value. They knew that life is too short to live without dreams. That there is no point to life if you don't find things you love.

WHAT IS MY PURPOSE?

The first job I got as a coach was exciting. Finally, I got to do this thing that I had set my intention on. But I was also a little bit insecure. The person I was coaching was much more successful than I was. He had started businesses, had black belts, and had been an Olympic athlete—yet he had come to me for coaching. What was I supposed to do to help him? He had already done so many incredible things.

But he came to me, he said, "Brian, I can't figure out my purpose."

It changed my perspective forever. Since then, I've found time and again that even people who have accomplished great things often haven't found a purpose. You may be able to achieve goals with serious effort, just like the pendulums that Zeland talks about want you to. However, the effortless flow and magnified wonder of finding your purpose is where the real magic happens. Once you find your true purpose, even the most successful person will move into a life flow that is filled with awe and wonder. This single step is transformational beyond anything else. It gives us what we need to go to that next level in our businesses, relationships, and lives.

People thought Steve Jobs was crazy for quitting college. He wasn't just quitting college, though—he was quitting what people thought he should do and instead chose what *he* felt he should do. He was answering an inner calling of destiny and purpose, which is the same calling that lies within each of us. The trouble is that we often overthink purpose. We think about what we want rather than doing things that give our lives the most excitement and meaning.

You might not believe that you have a calling yet. That's okay. Part of life is learning. Our experiences can be a contrast that show us what we want and don't want, yet "What do I want?" is often the very last question we ask ourselves. We look for purpose in what our parents, friends, schools, and communities say. We look at someone else and try to emulate them. We hang onto high school concepts of popularity that dictate how we dress, act, and work.

None of that is purpose. None of that answers your inner calling. No one else can decide what you're going to spend your time on. They aren't you.

You're not here to live Justin Bieber's life or your parent's life or your friend's life. You're not here to live the life that people tell you is smart or safe. You're not here to live a life that looks good or normal. Those are distractions that can take you away from your wonderful future.

The first step to designing your dream life is to block those distractions—to silence the voice that creeps in and says, "What if people think I'm crazy?"

We've all heard that voice. I've held back when teaching this very topic for fear of sounding crazy. When I finally started to think about it, I asked myself who exactly would think I'm crazy, and if it really mattered if they did. Who cares if people think you're nuts? It's better to be strange to others than to be a stranger to yourself.

"To be or not to be" is not a question but a choice that we all have to make. When you pursue what you love—when you choose to *be*—life becomes much more than a collection of ordinary experiences, choices, routines, and careers. None of my degrees, training certifications, or qualifications led me to my purpose. Many of the greats never even attended or finished school. This is something that comes from within, that you discover, choose, and pursue despite all external forces.

Purpose is the life path that will give you meaning and joy behind what you do. Acknowledge that your dreams and purpose matter—that to cease dreaming is to cease living. Do what you love. Follow your purpose like a path, and success will come.

WHAT *IS* PURPOSE?

In spite of the way most of us seek perfection, life is not a math test. There are no straight A's. There's also no failing grade. That kind of pursuit of perfection keeps us from purpose. What most of us wish for as a good life is actually a life warmed by mistakes and humility. A life touched with grace is impossible to live without occasionally picking your nose or stumbling while dancing in a crowd. It's in the mess that we find our message—we learn what we want by experiencing what we don't want.

Finding purpose is less like a test and more like misplacing your wallet only to find it was in your pocket the whole time. We spend our time desperately looking into the world for our purpose-driven life, when it was really right there with us all along.

Jack Canfield suggests asking yourself the question,

"What makes you feel like yourself?" Spend some time answering that question, not so much by thinking about it but by feeling it. Time travel to the future in your mind and feel what it would be like to live in this dream. Act it out around the house, if that helps you, then describe in two sentences what your ideal life would look like in one year. Then come back and describe that feeling in two sentences. Write it down.

Let go of perfection in this exercise. No one knows exactly what they are doing.

Two great names in film come to mind—Robert Rodrigues, who directed all the *Spy Kids* movies, and Sylvester Stallone. Neither of them had the luxury of perfection, training, or even support.

In *Rebel Without a Crew*, Robert talks about teaching himself to edit films when no one else would teach him. He raised the money to film *El Mariachi* entirely on his own, by giving blood and participating in an experimental research study. The same Hollywood that shunned him wound up loving his film—not because he was "touched" or because the circumstances were perfect, but because he followed his truth.

Sometimes perfection does show up but isn't aligned with purpose. When Stallone wrote *Rocky*, he had no job and his girlfriend broke up with him. He sold his dog at the local liquor store just to have money to live. In this state of desperation, someone offered him $100,000 for his script. The catch was that he wouldn't get to play

Rocky. Because his purpose was to be an actor, especially for the fulfillment of this role, he said no. The offers kept coming, up to a million dollars, and he kept denying them.

Finally, they offered $50,000 and the role of Rocky. And he took it.

So when you ask yourself what makes you feel like yourself—consider what you'd be willing to sacrifice to feel that way. It's a huge question, but much more manageable than contemplating the cosmic significance of your life. Thinking about the grand scheme of things can keep us locked on the couch, eating Doritos, and feeling overwhelmed. This question asks us to get outside and have experiences that make us feel like ourselves.

MONEY REALLY CAN'T BUY HAPPINESS

I believe the universe will help us find our purpose if we set out to look for it. But a lot of time, when people tell me they have found their purpose, it's tied directly to wealth. There's an enormous amount of unexamined assumptions underlying the pursuit of wealth and material goods, not the least of which is the assumption that these things will make you happy.

Check in with yourself. The last time you got a raise or received a big chunk of money, how did you feel? Good, I'm sure—but how long did that feeling last? Once the feeling goes away and everything returns to normal, we tell ourselves that it's because the amount wasn't big enough. If it had been the lottery or a big sale, *then* it would last. That's what we're led to believe. In fact, most companies, states, and countries need us to believe that. They use our drive to earn and the proceeds from lottery tickets to balance their budgets.

But sustained happiness does not come from money or the things it buys. It comes from purpose and meaning. It comes from developing a positive impact on the lives of others. Your life's purpose is not about "what" but "why." It's not what you would like to do or what you would like to make, but why you're here. Who you're meant to be. It's a question of design and meaning, not results.

HOW TO FIND YOUR PATH

In nature, it seems like everything has a definite purpose, from a place in the ecosystem to much simpler meanings. Holding a banana in your hand, it seems like the peel's purpose is to not get our hands messy. Even if it has a larger purpose, that's one we can't ignore. Do we have multiple purposes as well? Did the banana choose that purpose, and do we get to choose our own?

What if all of the problems we have faced were gifts given to us for a reason? Perhaps understanding our purpose can help us overcome those problems, because we'll see how they became advantages to us.

These are huge concepts to consider, and there are two types of methods to come to answers: indirectly and directly. Indirect methods use available evidence to construct a picture of life's purpose. It's like a crime scene investigation, where we use clues and come to a rough idea of what's happening. It may be difficult to get an exact sense of purpose, but it can get us closer than not asking at all.

Keeping the same metaphor, direct access methods would be more like an interrogation. This is communing directly with your soul to ask what your purpose is. It might be simply asking yourself, "What is my purpose?" every day, writing the answers down and tracking the patterns. It might be paying attention to the feeling that you get as you take actions. Your soul gives positive or negative feedback, you get a sense of fulfillment, passion, and making a difference. Things start to come easily. Chance and coincidence seem to support your goals and efforts. You notice a sense of synchronicity. Your efforts produce results, seemingly with little effort. You reach a flow state more—when you lose track of time and it's suddenly four in the morning but the thing you're working on is so exciting that you don't want to eat.

These are not always signals of purpose. My kids look like they're in a flow state when they play *Fortnite*, but I don't think that's their purpose. You might not think *Fortnite* is your purpose either, but sometimes it helps to get input from other people. You might hire a coach who can sense things in you and help give you feedback as well. Ask them what happens when you come into a room—what shows up when you show up? What do other people look up to you for? Write it all down—your deep beliefs and convictions about yourself, your gifts, and what is possible.

As you begin to discover your purpose, shape it into a statement. Write it down. Read it to yourself as you look in the mirror, every morning and every night.[42] Do it even when it's uncomfortable, because you deserve to tap in to your gift.

The greatest tragedy is that most people die without ever tapping in to their gifts. They never share their purpose with the world, never take the opportunity to make the world a better place. Realizing how close I came to that myself was deeply upsetting. All of the things I had learned and could do were empty without utilizing them toward my purpose.

Things might be tough right now, but when you find your purpose, it can change everything.

QUESTIONS TO PONDER

Consider the following questions. You might want to write the answers down in a journal, or perhaps just as a thought exercise. The logistics don't matter as much as the answers themselves.

---

[42] This is an exercise from Napoleon Hill's *Think and Grow Rich*, which I highly recommend.

- When have you felt most fulfilled?
- When has your life held the most meaning?
- When have you felt most aligned?
- When you were younger, what did you want to do when you grew up? What were you passionate about?
- What experiences in your life have prepared you for something incredible?
- What are you consistently drawn to, and why are you drawn to that thing?
- If you had a year to live, what would you spend your time doing?
- If you had a month to live—healthy, but limited in time—what would you spend your time doing?
- If you lived to be 120 and could look back over a long and fulfilling life with satisfaction, what would you remember the most?
- If you won the lottery and all your financial needs were handled, and you had already spent time buying toys and traveling and having fun—what would you do when you got bored?
- What would you do if your resources could go toward something meaningful instead of just survival?
- Imagine that all of your issues and wounds were things that your soul deliberately chose to develop and sharpen you. What have they given you?

DISTRACTIONS AND ROADBLOCKS

Many of us create an expectation that our purpose will appear out of nowhere. One day in the future, it will come in the form of a single word or a directive statement, and that will be it. We blow this singular future event so far out of proportion that the idea of purpose becomes overwhelming and not something that we can pursue. The reality is that purpose is a journey, and holding onto this picture of a singular event is one of the biggest problems people face in discovering their purpose.

Think about the moon missions that culminated with first steps on the moon. After the astronauts who landed on the moon came back to earth, they didn't know what to do next. The sheer sadness of losing this great purpose was troubling—you could hear it in their interviews.

Your assignment on this Earth is not just one event in the distant future that you hope you'll reach one day. We are in a continuous timeline of events that are constantly unfolding before us, and we are being propelled, moment by moment, on a timeline. Life-altering moments are going to happen. Bigger pieces of the puzzle will be revealed from time to time. The key is constant discovery. As you follow your vision as it develops over time, you will encounter destiny in big and small ways. The universe will begin to talk to you through coincidence and synchronicities, and your purpose will begin to take shape. It is the path that you travel, not the place where you arrive.

Do not tie your journey up into one thing. Allow the

discovery of your purpose to be ongoing and adaptable. Along the way, watch out for things that can pull you away from your journey.

PROCRASTINATION

Procrastination is a deadly disease that permeates the world we live in. Quite simply, the reason you procrastinate from day to day is because those tasks don't mean enough to you. The reason you just go through the motions in life is because you aren't aiming at anything worth the effort.

In other words, procrastination is a sign that you're not on the right path.

We like to complicate things like this, but it's quite simple. Ninety percent of the time, when someone comes to me struggling with procrastination, some kind of social media or online addiction has them distracted. They're on Facebook or Instagram four or five hours every day, and it has taken them from their purpose.

Now, if you really believe you will find your purpose on social media, then that's fine. But if it's just helping you to procrastinate, it may be a sign that you need to overcome the addiction and get back on your path.

LIMITING BELIEFS

The classic line that everyone says is, "I can't achieve my

purpose because I don't have the means." Sometimes, we will take it a step further and just say, "I'm not good enough." Maybe you don't have the means right now, but hanging onto that belief becomes a self-fulfilling prophecy. Because you don't have the means, you sit around for four hours watching *Game of Thrones* when you could be working on that software or creating that piece of art. If you never aim at the bullseye, you'll only ever hit random targets. Just by putting those things in our minds, we stop ourselves from expanding and following the path of purpose.

The beliefs that we create from our younger years onward define how we go about maneuvering our purpose later. Whether they are beliefs about fitting in or being good enough, or more complex beliefs rooted in difficult circumstances, they are limiting.

Even if you still fully believe those limiting beliefs, bringing them out into the open and acknowledging them, usually in the form of a journal, can help eliminate them. Try going for a couple of days without believing those things. Try it on, like a brand new coat. Picture them in a box outside of your home, and they'll be there for you to pick them up again later if you need to.

You'll find it's much better to live without them. Those beliefs keep you from finding your purpose, because they tell you that you can't, won't, shouldn't, or will never do the thing that makes you feel most like yourself. The thing that gives you meaning and keeps you alive. Let them go, and you'll find the path much easier to walk.

PERSONAL ASSOCIATIONS

You end up being defined by the five people you spend most of your time with. It's true for behaviors and finances as well—I would almost guarantee that you have the average income of the people you're surrounded with. This isn't just self-help speak: we actually start to develop genetic similarities with the people we're with all of our lives. If you've noticed, old married couples start to even look alike.

These people who we're surrounded with begin to shape our purpose as well. They become the little voices in our head. Part of me didn't want to start my podcast or write some of my books because certain friends would think I was crazy. Sometimes you don't have a choice about the people in your life, but other times you can let those people go.

Tim Ferriss talks about associations in *The 4-Hour Workweek*. We need to be strict about who we let in as mentors and close friends. If there are people who aren't moving you closer to your true purpose, create some space between you and them. You don't have to cut them out of your life, but you can begin to "politely not answer" or move them out of that inner circle that has so much influence on your life and on finding your purpose.

INSECURITY

We all have that friend who never goes out or is too shy

to move beyond their insecurity. That friend might think they are too fat or too ugly or a host of other things.

Let me be really clear on this one: their insecurity is wrong, as is yours.

I don't care what that insecurity is. I don't care how ugly you are, how fat you are, how unpopular or unsuccessful you are. None of it matters. Once you start following all of the paths possible to you, those minor details won't matter. You can still shine. Don't let your insecurity pull you from your purpose.

LACK OF ACCOUNTABILITY

A lot of times, we don't find our purpose because we're only accountable to ourselves. If we say that we're going to work out six times in a week and then don't, there are little voices in our head telling us it's not a big deal. It's only when we get a friend or coach on board to help us stay accountable that the voice can be silenced. We need another person who will tell us, "Hey, you should've worked out."

In the absence of a person, there are generic apps that can fill the gap. You can record your reps there or list out your goals. But in my experience, having a person that you're accountable to is a gamechanger. You cannot delete or ignore a person, and the real accountability that provides can subtly shift your habits and processes. It's for this reason that coaching works.

FEAR OF FAILURE

Fear is a killer, and the fear of failure might be the strongest of them all. I think that, in an alternate timeline where nothing had happened to make me question my mortality, the fear of failure would have dominated my life.

We are so afraid of failure that we often let it keep us from taking action at all. I have friends who are geniuses but have never submitted their books or stepped out into the audition because they are afraid of failing or being embarrassed. We don't want to do things because there is some small chance that we will fail. The catch is that you probably will, but that it's a good thing. Every successful person has built a mountain of failures. Failing is a wonderful part of your journey.

Don't let a fear of failing be the reason that you never found your purpose or the life that you should be living.

DESIGN YOUR DREAM LIFE

Probably the most powerful way to access your purpose is to read. Books connect us to people we'll never know, in times and places in history that we'll never travel, to ideas and constructs that we'll never experience. By reading fiction and nonfiction, autobiographies and memoirs, we can literally live out entire timelines. We see how other people found their purpose and follow their informational path.

If you're lost and don't know what your purpose is, this is the very best way to find it. There are so many books that speak to your journey, no matter what your journey is. Grab onto their timelines—they're available and waiting for you as giant archives of different pasts and futures.

You might also tell your own story. I've left journaling prompts wherever I can in this book because I believe in its power. When you write down your story, you gain a better understanding of what you've been through. If not a journal, then a blog. A poem. A podcast. Simply by going through the process of telling your story, you'll gain a greater understanding of yourself. You'll see things you might have missed before.

Most importantly, you'll regain ownership of the story that you're telling with your life. When people come to me looking for their purpose, they immediately try to gain control of their story by firing their boss and setting out on their own. In my experience, the internet has made these goals so much more accessible and achievable than ever before. We can take passions and side hobbies and turn them into businesses.

Just keep in mind that your dream job is not your purpose—though your purpose can be integrated into your job. The vast majority of us will never find the work, income, or circumstances that provide unending fulfillment. The goal is to eliminate what's holding us back and seek out what will propel us into the future that we want.

Design the life that you want. Look for the opportunities that align with your purpose and your passion. And begin to maneuver into the reality that maximizes it all.

> **JOURNALING PROMPTS**
>
> *What am I willing to put up with?* People who have found their purpose are often people who were willing to put up with a lot of less-than-ideal situations. Life is not going to be rosy and wonderful, even when you're aligned with a purpose. Sometimes you have to work hard. Sometimes there is a sacrifice—some sort of cost or struggle—that you'll have to face. Knowing this at the onset can help you to narrow down your purpose. For example, if you want to be a tech entrepreneur but you can't handle failure, you're not going to make it very far. On the other hand, if you're willing to work eighteen-hour days for a couple of months to get your book written or your software complete and that sacrifice seems worth it to you, then you'll start moving closer to your purpose.
>
> *What would a younger me think about me now?* Often, the wisest version of ourselves was when we were seven years old. Ask that kid what they think. After all, this is the life that your younger self wound up living. What would they say about it? Creating that perspective can give you some good answers about who you are and where you're going. (Science fiction

writer Steven Barnes uses this technique every morning upon awakening.[43])

What can you do to help the world? Wanting the new Lamborghini or that perfect mansion is understandable and fine, but thinking outside of yourself is often transformative in a way that wanting things can never be. When you really find your purpose—when everything starts to make sense—it's often when you're thinking about how you can help other people with your talents and skills. Your heart will line up with your mind and soul in many ways.

If you knew you were going to die tomorrow, how would you want to be remembered? It's okay to think of purpose as how you want to be remembered. That sounds like a contradiction against not worrying about embarrassment, but it's not. The difference is that when we have found our purpose, we control the way we are remembered. We decide how we are perceived by the world. Whether or not they like it is up to them—the important thing is that you've done what you loved to do.

---

43 Brian Scott, "Interview with Steven Barnes: Writer, Life Coach, Expert in Martial Arts, Tai Chi, and Yoga." Podcast Audio. *The Reality Revolution Podcast.* July 1, 2019. http://www.therealityrevolution.com/interview-with-steven-barnes-writer-life-coach-expert-in-martial-arts-tai-chi-and-yoga-ep-70/.

PART III

A NEW REVOLUTION

CHAPTER EIGHT

THE HEALTH REVOLUTION

It is health that is real wealth, and not pieces of gold and silver.
—MAHATMA GANDHI

Before I started writing this chapter, at breakfast one morning I got suddenly, unpredictably sick. I felt extreme pain, my skin turned cold, and my face broke out in a sweat. I tried to fall back on the techniques that I have learned and taught to others, but I thought I was dying. Nothing was working, and at the hospital they told me I had a kidney stone.

Later, lying on my side in the hospital bed clutching a heating pad, I started to use breath and visualization. For moments at a time, I could take away the pain completely. In that moment, I realized that as much as I knew about realities, I still needed to learn more about my health.

I began to consider the implications—what if, at some point in the future, we were to discover the secret to living forever? Not through some medicine that we could take or a procedure we could have, but with an understanding of the mind-body relationship. Imagine that world. Imagine us discovering that the placebo effect is really just a quantum jump into a body that's already healthy. Imagine science being so advanced that we learn how to regularly interact with our bodies on a quantum level.

To take it a step further—what if this is already accessible, but we're so resistant to changing our realities that the part of us that doesn't believe pulls us back into sick bodies?

A lot of us come into an understanding of reality creation because we want to create a great job or manifest money, but the fundamental thing that we are creating at all times is our own body.

Every time we heal, we shed our skin and become something now. Every single exhale is another wish into the universe that we're wasting. Every thought recreates part of your body. Our bodies are constantly in motion, constantly being created by the food that we eat and the thoughts that we entertain. Bringing health onto our list of intentions instead of taking it for granted will help us better create the universe around us.

Our future extends beyond anything written in science fiction, and that includes an understanding of the human body that is about to explode in the way that technology has in recent years. Already, biological processes from photosynthesis to bird migration to the sense of smell and the origins of life are increasingly understood to be quantum processes, which means the same quantum mechanics we study for reality shifting affect our very DNA. We're finding that chromosomes possess wave dualities, just like we saw in that classic double-slit experiment, as well as holographic memory.[44] The placebo response has become so powerful that it affects research, and medical journals continue to catalogue cases of individuals with spontaneous remissions and miraculous recoveries.

The truth is that we don't yet know how to live forever. We aren't familiar enough with our cells, memory, and quantum selves to reliably manipulate our health. But there is something fundamentally powerful in the connection between health and reality that we need to explore. Put simply, the reality revolution is beginning to happen already. Human potential is vast and untapped, with an extraordinary capacity to influence the world, right here at our disposal.

THE POWER OF CELL MEMORY

You might have people in your own life who have "beaten the odds." It happens more than we realize, because as a

---

[44] Richard Jorgenson, "Epigenetics: Biology's Quantum Mechanics." *Frontiers in Plant Science.* 2, 10 (2011). https://www.ncbi.nlm.nih.gov/pmc/articles/PMC3355681/.

society we're too quick to dismiss them as an abnormal occurrence. No one looks deeply into these or thinks of them as reproducible occurrences. Instead, we treat them as something bizarre or unusual and leave it at that.[45]

I believe we need to reframe our understanding of biology in more miraculous terms.

Most simplistically, think about the first time you skinned your knee and your mother kissed it to make it better. What if that's the first true reality shift that we experience? All those times when we immediately forgot the pain, whether it was from a placebo or a comforting parent, I believe it's because part of us has attached to a healthier reality. I believe things like the placebo effect, the immune responses, and the ability to forget the way something felt are all happening on a quantum level.

Mark Chenoweth comes to mind as a more direct example. He was born with spina bifida, a crippling disease that left him unable to walk. In 1998, he consulted with his doctor about taking scuba diving lessons. Even though the doctor forbade it, Mark took a vacation to Menorca and persuaded a diving center to give him lessons. After diving to fifty-five feet, he surfaced to find that he could walk again. Three days later, his legs lost sensation again, and he immediately went back out to dive. After a while, he noticed that the deeper he went,

---

45   There is some research happening. Dr. Joe Dispenza has hooked people up to EEG machines and tested heart coherence in large groups of people, and in multiple cases he has documented sudden remissions.

the longer he could walk afterward. No one has been able to explain why this happened.[46]

Though it is just a theory, I believe that his infinite self gained access to a parallel universe where the pressure was different and he could walk, and recreating that environment helped his body remember that reality. He had to continually recreate those circumstances to keep recalling that memory.

This is my explanation for his experience, because I believe there is a part of our subconscious that remembers all of our lives in all realities. We do seem to have evidence that the personality resides in the cells of the body as much as it is in the brain. It's called "cellular memory" and is a point of research that indicates organs like the heart and hands contain their own memories. Signals of this possibility include the story of William Sheridan, who inexplicably developed a passion for art and began to create beautiful sketches during recovery from a heart transplant. Later, he found out that his donor had been an artist.[47]

The mind-body connection is powerful and real, and the conscious mind is not in charge of that connection. While thoughts are the language of the brain, feelings are the language of the body. Managing how you think and

---

46 "Spina Bifida Man Walks Again," Possiblemind.co.uk, June 2013. http://possiblemind.co.uk/spina-bifida-man-walks-again/.

47 Dr. Paul Purcell has done specific research around this phenomenon. You can read more in his study, "Changes in Heart Transplant Recipients That Parallel the Personalities of Their Donor," *Journal of Near-Death Studies*, (2002) 20: 191. https://doi.org/10.1023/A:1013009425905.

feel creates a state of being where your mind and body work together. If we learn how to overcome the loop of feelings, we can use this level of the reality revolution to change the body.

RETRAINING THE BODY AND MIND

We have already accepted that we have the power to improve the quality of our air and water, crime and accidents, and even the education levels of our children using the power of our minds—but we also may be able to live longer and affect our own health and the health of the people around us.

Feelings are the modus operandi of the body, and the emotions that you continuously feel will be memorized, equal to the unconscious, hardwired mind. Every time you have a thought, in addition to making neurotransmitters, your brain makes small proteins called neuropeptides that send a message to the body, which then reacts by having a feeling. On the other side, when the brain realizes that the body is having a feeling, it generates a thought to match it.

This creates a loop of thinking and feeling, and when it operates long enough it starts to become a memory in the body. The cells and tissues receive chemical signals at specific receptor sites, like docking stations where chemical messengers fit in like puzzle pieces. Those docking points alter the body physically and train it to memorize the loop. Attempting to break the loop often only strengthens it with stronger thoughts and feelings.

Every time you sit and think about that thing, you have the same feeling. When you have that feeling, the body doesn't know it's not still experiencing it. As the loop integrates into your normal life, you might start to have the feeling without knowing why. Eventually, your body won't know what to do with that energy and will begin to send it to other parts of the body, which can make you sick. As the same genes become repeatedly activated over and over again by this loop of information and feeling, they begin to wear out like gears in a car. The proteins that the body creates become weaker and with lesser functions, such as illness or aging.

In time, the intelligence of the cell membrane that is constantly receiving the same information, can adapt to the body's needs and demands by modifying its receptor sites to accommodate more of these chemicals. The body becomes addicted to the feelings loop, biologically conditioned for that environment. Just like a substance addiction, a cell that is bombarded will become desensitized and require more chemicals to activate. This requires more emotional memories and unconscious habituation. It becomes so ingrained that it defines your mind and body.

Often, we are making decisions based on these states, even before we consciously make the decision itself. If your conscious mind, which has about 5 percent control, tries to change this process, it's swimming upstream against the 95 percent unconscious control. By the time you're an adult—thirty, forty years old—you're just a set of programs in the body.

The only way to affect real change is to actively, fundamentally change the way that you think.

GUIDED MEDITATION

Go back to a time in your life when you had perfect health. I call it the bliss body. Was it when you were a kid? Was it when you were in college or working out regularly?

What did that body feel like? Memorize those points in time when your body felt good, pleasurable, happy, and energetic. Anchor those feelings and bring them back to where you are now.

Try my bliss-body meditation and bliss-body sleep meditation.[48][49]

---

48 Brian Scott. "Guided Meditation: The Blissbody Meditation." Podcast Audio. *The Reality Revolution Podcast*. May 2, 2019. http://www.therealityrevolution.com/guided-meditation-the-blissbody-meditation-ep-36/.

49 Brian Scott. "Deep Sleep Meditation: Regenerating in the Source Field - Blissbody Sleep Rejuvenation." Podcast Audio. *The Reality Revolution Podcast*. June 1, 2019. http://www.therealityrevolution.com/deep-sleep-meditation-regenerating-in-the-sourcefield-blissbody-sleep-rejuvenation/.

REMOVING THE LIMITS

Healthy people still get sick. You can exercise every day and eat well, yet still get sick if your thoughts aren't optimized for health. If you don't have a specific goal or desire, for example, I believe your life force is weakened and your mind is driven into a box.

One of the hardest things to witness is someone who has lost hope in life. Maybe they have experienced something tragic or have reached a certain age, but giving up can be fatal. If you want it to, your body will simply die for you, like the Holocaust victims we noted in chapter 7. The first step to longevity, then is to find hope and a purpose.

Biologists are currently honing tools for deleting, replacing, and editing DNA, such as the CRISPR technique. Using a modified bacterial protein and RNA, they are able to change the DNA of an organism while it's still alive.[50] If we live long enough, technologies like this will develop until we can live forever. By 2015, some one thousand research papers had been published that mentioned CRISPR. The singularity is coming, and energy and the subconscious will be necessary parts of health when it does.

In the coming century, most of our conceptions of life will be turned on their head. Belief in the inevitability of death will change. Maybe through some technique or mechanism, or maybe through quantum computing or

---

50 Ray Kurzweil, *Fantastic Voyage: Live Long Enough to Live Forever* (Plume, 2005).

simply quantum jumps—the ability to create massive, fast transformations is around the corner.

Many experts believe that, within the decade, we'll start adding more than a year to human life expectancy every year. Aubrey de Gray believes that we will successfully stop aging in mice within the next decade, and then human therapies that halt and reverse aging will follow five to ten years after that. At that point, with each passing year, the end of your life will move further and further into the future.

I meet people in their sixties who have given up. They think they only have a short time left to live. Please know that there is hope. It is my serious intention to live forever. There is no reason that we can't change the way we think about our bodies and then live long enough to reach that point in history. Yes, there are degenerative processes in our bodies, like kidney stones and emotions, but that's no reason to give up.

Forget about the past. Imagine that each of the hundred years in the twentieth century is the equivalent of ten years from now, and that time is shortening as we go. Soon, twenty years will go by in what we know as a year. Our progression will expand and compound. Soon, we will have nanotechnology that will enable us to rebuild those degenerative processes. DNA printing will create any product that we need, and technology like CRISPR will edit our genetic code directly. What we think we are doomed to get—cancer or Parkinson's or Alzheimer's—won't have to happen.

We won't be stuck with our DNA anymore, and it's possible that we won't be stuck with our brains and bodies, either. Perhaps we will be able to upload our consciousness to the internet. A company called Humai intends to resurrect humans within the next thirty years, transferring consciousness into a robot body.[51] They will not be the first. When the technology is fully developed, there is a chance that death will become a transfer into an artificial body—or perhaps into another fully functional, healthy human body just like the Netflix series *Altered Carbon* (where consciousness is transferred by plait, of course).

This might all sound bizarre, and maybe it is irrational. But it is a potential that is absolutely worth considering.

RETHINKING MEDICINE

The reality revolution is changing the way we look at medicine, most notably with a resurgence in integrative medicine. For example, if you go to a mainstream doctor and tell them you have a rash on your arm, they will tell you, "I'm sorry to hear that. Take this cream and rub it on the rash," and send you on your way.

They aren't thinking about the cause of the rash or what it might indicate is happening internally. They're going after the symptom, not the cause. Integrative med-

---

[51] David Gershgorn and Sarah Felt, "Humai Wants to Resurrect Humans within 30 Years." *Popular Science*. November 23, 2015. https://www.popsci.com/humai-wants-to-resurrect-humans-within-30-years.

icine, on the other hand, is a partnership, with the main purpose being the promotion of a self-healing capacity.

More and more, the body and mind are being integrated into medicine, which looks like using diet, exercise, meditation, and stress management to potentially affect genes. According to the *Wall Street Journal* in 2009, studies are uncovering our bodies' remarkable capacity to begin healing through these methods—and much more quickly than we once realized.[52] These lifestyle factors can address chronic diseases such as diabetes, heart disease, and cancer, in documentable ways that change health and wellbeing.

Beyond a belief in self-healing, integrated medicine stresses the importance of the mind-body connection. It's about a healthier lifestyle, in spite of the fact that we are not ever guaranteed a clear miracle.

THE MENTAL PLACEBO

Whether you've been trying to affect positive change to create a new state of being or you've been running on autopilot, you have always been your own healing placebo. The brain's power over the body is undeniable. According to the National Cancer Institute, people who are undergoing chemotherapy get sick when they smell the treatment, even before the actual treatment begins. Eleven percent

---

[52] Deepak Chopra, et al, "Alternative' Medicine Is Mainstream," *Wall Street Journal*, 2009. https://www.wsj.com/articles/SB123146318996466585.

of people feel so sick before treatments that they vomit. Others feel nauseated in the car on the way to the hospital, and still others throw up in the waiting rooms. It has been termed "anticipatory nausea," and we might relate to it even without having had chemotherapy. Most of us have gotten sick when thinking about something.

Imagine then the power that our thoughts have over our health. How much other sickness is happening from our own minds? Could a single thought that a person accepts make them physically better or worse?

My dad, my grandmother, and my grandfather all had Parkinson's disease, so I'm always reading about the studies around it. A University of British Columbia in Vancouver study assessed a group of patients with Parkinson's, who were meant to receive a drug that would significantly alter and improve their systems. Instead, they were given a placebo.

Half of the patients who received the placebo displayed much better motor control—and it was more than an external change. Researchers scanned the patients' brains to get a better idea of what happened, and in those who responded positively, dopamine levels had increased to as much as 200 percent more than before the injection. To get the equivalent of the dopamine increase they experienced, they would have needed a full dose of amphetamine or a mood-elevating drug.[53]

---

[53] R. de la Fuente-Fernandez, et al, "Expectation and Dopamine Release: Mechanism of the Placebo Effect in Parkinson's Disease," *Science* 293(5532):1164-6. 2001).

Their brains had become the perfect pharmacological source for the relief that they needed.

THE POWER OF POSITIVITY

We can see this evident in a number of studies indicating healing-linked proteins are suppressed and released based on an emotional state. In one study, after couples discussed previous disagreements, this suppression rose to about 40 percent when the discussion ballooned into "significant conflict, sarcastic comments, criticism, and put-downs."[54] In a reverse effect, positive emotions and reduced stress trigger epigenetic changes that improve health.

Researchers at the Benson-Henry Institute for Mind-Body Medicine in Boston looked at the effects of meditation, which is now known for eliciting peaceful, even blissful states of gene expression.[55] In 2008, volunteers were trained for eight weeks around mind-body practices including meditation, yoga, and repetitive prayer known to induce a physiological state of deep rest. They also followed nineteen long-term daily practitioners of those same techniques, and at the end of the study period, novices showed a change in 1,561 genes, with 847 upregu-

---

54 Lian Block, Claudia Haase, Robert W. Levenson. "Emotion regulation predicts marital satisfaction: More than a wives' tale." *Emotion*. Feb; 14, 1 (2014): 130-144. Published online November 4, 2013.

55 Sara Martin, "The Power of the Relaxation Response." *Monitor Staff.* 39, 9 (2008): 32. https://www.apa.org/monitor/2008/10/relaxation

lated for health and 687 downregulated for stress, as well as reduced blood pressure and reduced heart rate and respiration rates. The experienced practitioners expressed 2,209 new genes, with most of the changes involving an improved response to chronic psychological stress.[56]

In a second study in 2013, the same researchers found that just one session of meditation in novices and experienced practitioners alike produced changes in gene expression.[57] Long-term practitioners, again, derived more benefits than the novices. The genes upregulated in this session included immune function, energy, and metabolism, while the genes downregulated included inflammation and stress.

Before either of these studies, the Mayo Clinic in 2002 published findings on 447 people over more than thirty years were physically and mentally healthier if they were optimistic than not. According to the study, optimism means "best," suggesting that the people who were healthier focused their attention on the best future scenario. Optimists had fewer problems with daily activities as a result of their health or emotional state, felt more energetic, experienced less pain, and had an easier time with social activities. They felt happier, calmer, and more peaceful for more of the time.

---

[56] JA Dusek, et al, "Genomic counter-stress changes induced by the relaxation response," *PLoS One*, 2;3,7 (2008):e2576. doi: 10.1371/journal.pone.0002576.

[57] Ivana Buric et al. "Frontiers in immunology. "What Is the Molecular Signature of Mind-Body Interventions? A Systematic Review of Gene Expression Changes Induced by Meditation and Related Practices." *Frontiers in Immunology. June 16, 8 (2017): 670.*

In another Mayo Clinic study, 800 people followed for thirty years showed optimists living longer than pessimists. Researchers at Yale, following 660 people aged fifty or older for up to twenty-three years, found that those with positive attitudes about aging lived more than seven years longer than those with a negative outlook about growing older. Attitude has more of an influence on longevity than blood pressure, cholesterol levels, smoking, body weight, or level of exercise.[58] Other studies have looked more specifically at how attitude affects heart health, such as Duke University's study of 866 heart patients—those who felt more positive emotions had a 20 percent greater chance of being alive eleven years later than those who habitually experienced negative emotions.[59]

Perhaps the most striking results come from a study of 255 medical students at the Medical College of Georgia, followed for twenty-five years. Those who were the most hostile had a five-times greater incidence of coronary heart disease.[60] Another from Johns Hopkins in 2001 showed that a positive outlook may offer the strongest

---

[58] Levy, Slade, Kunkel, and Kasl. "Longevity Increased by Positive Self-Perceptions of Aging." *Journal of Personality and Social Psychology.* 83, 2 (2002). 261-270

[59] I.C. Siegler, P. T. Costa, B. H. Brummett, et al., "Patterns of Change in Hostility from College to Midlife in the UNC Alumni Heart Study Predict High-Risk Status," *Psychosomatic Medicine,* 65, 5 (2003): 738-745.

[60] J.C. Barefoot, W. G. Dahlstrom, and R. B. Williams, Jr., "Hostility, CHD Incidence, and Total Mortality: A 25-Year Follow-Up Study of 255 Physicians," *Psychosomatic Medicine,* 45, 1 (1983): 59-63.

known protection against heart disease in adults who are at risk due to family history.[61]

This is a matter of life and death. Positivity can overcome your genetic history, add years to your life, and give you healthier years. Having the right attitude alone seems to work as well as or better than eating the proper diet, getting the right amount of exercise, and maintaining the ideal body weight. Perhaps a more universal explanation is that the people who are optimistic and positive are constantly shifting into a healthier reality.

UTILIZING THE MIND-BODY CONNECTION

After my kidney stone episode ended, I went in to research mode again. I read *Heal Your Body* by Louise Hayes, *The Language of Your Body* by Julia Cannon, and *The Emotion Code* by Dr. Robert Bradley. Each of them discusses the ways that we store emotions in the body and how that can make us sick. When we thought we didn't get scared or we were proud of holding our anger back, sometimes that is held, unreleased emotion that we aren't even aware of, but the body knows and responds to it.

Many practitioners say that you can use things like magnets or communication with your subconscious to

---

[61] D.M. Becker, L. R. Yanek, T. F. Moy, et al., "General Well-Being Is Strongly Protective Against Future Coronary Heart Disease Events in an Apparently Healthy High-Risk Population," Abstract #103966, presented at American Heart Association Scientific Sessions, Anaheim, CA, (November 12, 2001).

unlock emotions. I believe we need little more than the intention to release the energy.

Once old emotions are released, we need to get more comfortable expressing emotions as they arise. Most of us are good at expressing happiness, but few of us express things that make us scared or angry. Release discouragement, guilt, anger, and hopelessness, just to name a few.[62] If you have a sickness of some sort, ask yourself if it is something you're creating. Is it related to thoughts that you're having on a regular basis? Is it an emotional loop that you've created? The only way to avoid those feedback loops is to identify, express, and release emotion.

CHOOSING HEALTH

What happens when you wake up in the morning and feel that little tickle in your throat? Everyone says the same thing: *I'm going to get sick.* I just had this happen the other day, on a morning when I had a ton of work to do.

Instead of confirming the awareness of sickness, instead I said, "No," and I physically stepped to the side as if I were entering into another body. "I'm going to be fine," I said, and three or four hours of ignoring the symptoms later, I was fine.

If we can make ourselves sick, we can also make ourselves better.

---

[62] In *Heal Your Body*, organs and illnesses are listed with their real, metaphysical causes and affirmations to address them. In my case, it seems I had fear trapped in my kidney.

When I was a kid and didn't know any better, I remember riding my bike down a hill going away from my house. This was back when I was eleven or twelve, before cell phones but while we still went out to play in the neighborhoods without our parents. I hit the front brake and started to fall, and as I rolled around trying to avoid hitting my head, my ankle took the brunt of it. I swore it was broken, and there I was in the middle of nowhere.

I remember thinking I had to get home for dinner, so I got up and walked toward my bike telling myself, "My leg is not broken." I got on the bike and started pedaling home, just thinking about how I needed to get back. By the time I got home, my ankle was already better. Two days later, I didn't have any pain at all.

When I was dealing with those kidney stones, I used breath to focus my thoughts and dissolve the pain. This is because most pain begins purely as a mental/emotional creation before it manifests as a physical sensation. If we catch pain in the early stages or as "phantom pains," we can resolve it with the mind and spirit. Over time and with steadfast immersion, I believe it is possible to reverse or heal pain entirely.

The easiest pain to heal is psychosomatic pain that requires no medication. It usually rises in conjunction with fear, stress, or stuck emotions. Eventually, I believe we will be able to mitigate pain as well as reduce the need for anesthesia in some cases.

Of course, our bodies could be telling us something

when we get symptoms of illness or injury. The point is not to directly ignore them. Deal with illness and go to the doctor whenever you need to. But if you continue to focus on the sickness or the injury, you might not find the results that you're hoping for.

What's happening in these moments of instant healing? There's no way to know for sure. Go back to them. Find out what happened. That one access point may be the key to understanding our healing and how to make it happen in the future.

INGESTING WITH INTENTION

We have the most control over our health before we put anything at all into our bodies. We are what we think and are and digest, though there aren't as many nerve endings inside our bodies to signal warnings to us as there are externally. We have to rely on intuition to keep us in line.

Trust your instincts with every bite of food and drink of water. Remember the power that water has to carry intention—when you drink water, your thoughts are making an imprint on that water. If you're dwelling on something terrible, you're going to drink that emotion and disburse it into the world. This is true of food as well, since most foods have water in them. You're literally changing the structure of what you eat and drink before you ever get a bite.

Give yourself a brief second before every meal to have

gratitude, joy, and happiness. Put your intentions into your food and swallow them. This is a conscious level of eating that gives us the ability to become healthier with every breath, every bite, every drink. Not only is this powerful in practice, but it gives us a sense of ownership over our health.

DNA MANIFESTATION

One of the biggest things I hear about health is someone saying, "I've got bad genes" or "I'm not responsible for it." But even before CRISPR gets a hold of our DNA, the science of epigenetics tells us that how we behave affects our gene expression. We absolutely have the ability to overcome our genetic coding, or any other limiting belief we have about our health.

Bruce Lipton's book *Biology of Belief* talks about an epigenetic revolution that we're going through as a society. It explains epigenetics as a movie script for the DNA, and that the script can be adjusted. DNA is not a fixed, individual thing but something that constantly transmorphs and changes and activates, expressing itself differently over the course of your life. You get to be the director of that script.

While your body replaces its cells every seven years, some of them are replaced even faster than that. Red blood cells live only four months. White blood cells live just over a year. Skin cells make it two or three weeks, while colon cells survive for four days and sperm cells for three days. So the body is being replaced much faster than we realize, and DNA contains the raw information for its replacement. Then each of the cells manufactures proteins (from the Greek *proteios*, or *primary importance*) that construct structure, intricate functions, and complex interactions that make up our physiology.

To take that realization a step further, Dr. Peter Gary discovered powerful evidence that the source we talked about earlier in the book actually operates through our DNA. He put a sample of DNA in a small quartz container and then zapped it with a laser. Then, with sensitive equipment that could detect even a photon of light, he found that the DNA acted like a light sponge, or a mini black hole. Somehow, the DNA molecules pulled all of the photons of light into itself and stored it in the spiral. When the DNA was removed, it somehow held the light in the same place and structure. His work suggests that we might be comprised of more than just blueprints—we are leaving a trail of light everywhere we go.

Japanese researchers have found that the human body glimmers, with light fluctuating through the day in cycles.[63] The light is brightest in the afternoon and dimmest in the evening. Could it be that the light source

---

63 Masaki Kobayashi, Daisuke Kikuchi, and Hitoshi Okamura. "Imaging of Ultraweak Spontaneous Photon Emission from Human Body Displaying Diurnal Rhythm," *PLoS One*. July 4, 7 (2009).

is energy and we are constantly shedding and gaining it throughout the day? Is this light what makes us alive?

Imagine that: your DNA has the power of a black hole, strong enough to store and control visible light. The rational explanation is that there has to be an energy field paired up with the DNA molecule like an energetic duplicate or quantum entanglement. We are not interacting with DNA alone, but the source field that it carries.

Conversely, Professor Fritz-Albert Popp, a German theoretical biophysicist and genius of his time, found that carcinogens would absorb light and then scramble it into a different frequency. He got a student of his to build a device that could detect single photons of light, and they began researching other substances as well. He ultimately concluded that a biophoton field around living organisms contains information that is useful to that organism. We each have a field around us that affects us. When we're stressed, for example, we give out more light than usual as a sign that we need additional rejuvenation.

When Popps described this light, he noted that they are waves, just like a vibration or frequency. On a graph, it would look like a heartbeat. Dr. Rupert Sheldrake picked this subject up and called these waves morphic fields. After nine years of intensive study, he concluded that biochemistry could not provide the answers he was looking for, and began to draw on the work of French philosopher Henri Bergson. He ultimately proposed that memory is not stored in the brain at all, but in the morphic or biophoton field.[64]

It could be that the mind sending information to our feet is actually sending it by jumping right out into a field of light rather than traveling through the blood. It could be that the brain simply receives information, while your "hard drive" is located in the biofield around you.

Imagine if we understood how to interact with that field—and if that field were literally a transdimensional point between multiple universes that are constantly in flux. How would we think differently about our bodies and our health?

TOOLS FOR A HEALTH REVOLUTION

Be still. To be in good health, we clearly need to relax. For at least fifteen minutes a day, in some way, find a way to be still. Rest gives a boost in energy, increases a sense of calm, and creates happiness. It has a ripple effect on your overall body and is part of a meditative practice. Stillness gives you the ability to connect to yourself.

Set intentions. Lynne McTaggart's research on meditation indicated that intentions should always be specific and in the present tense.[65] If you have lower back pain,

---

[64] Niki Gratrix, "Supplementing with Light: Can Laser Nutrients 'Speak' to DNA?" *CAM MAGAZINE*, Nov (2012): 33-38

[65] Lynne McTaggart, *The Intention Experiment: Using Your Thoughts to Change Your Life and the World* (Atria Books, 2008).

the intention would be, "My lower back and sacrum *are* free of all pain *and now* move easily and fluidly." You can write these intentions down or create a collage of photos or magazine pictures, but be specific. If it's about a single part of a single organ, say so. The more specific the intentions, the more effective they were to help people heal. In addition to the intention, visualize yourself in that reality. Imagine a healthy you in this present moment. Feel the feeling of being pain free.

This is not an easy practice if you are in pain, as I learned when I had the kidney stone. The idea of creating that bliss body is to remember a perfect state. When you're happy, joyful, and feeling healthy, create an anchor in that moment. Pinch a part of your arm or look at a symbol. Later, when you do get sick, you can give yourself that reminder to bring the happier feelings back. It doesn't always work right away, but the more you do it, the more likely you will be to create a new loop.

Fast. A lot of times, when we eat or exercise or do things outside of our bodies, the message is, "I don't trust you to heal, so I'm going to do it for you." That's okay if you're following your intuition and doing the right thing, but don't forget that the body has incredible healing technology on its own, beyond anything you can imagine. The one thing you can trust is your body, and one way to express that trust is by fasting. This gives the body time to heal and do its job.

The effect that fasting creates is called *autophagy*,

and it helps the cells to repair, the digestive process to rest, and keep everything generally ticking nicely. You don't have to be extreme in your fasts, like Twitter's Jack Dorsey having only water every weekend. Simply start by giving yourself twelve hours every day without food. In the practice of intermittent fasting, you would stretch that period to eighteen hours and only eat during six.

Movement. I exercise five or six times each week now, and it has transformed my life. About thirteen years ago, I had let myself go. I weighed 320 pounds, and I didn't care what I was eating or how I was moving. When I decided it was time to lose weight, I changed my foods and picked up a P90X workout. Everyone is different and needs different things. I've learned that the human body constantly adapts, and there is no one program that will help everyone. In fact, a lot of exercise programs now build that concept in, understanding that the body is constantly adapting and getting used to what you're putting it through.

When we change our movement choices up on a regular basis, it shocks the body with new exercises—and it's not hard to do. Exercise is not about pumping iron or going to spin class. It can be about simple movement. Yet according to the WHO, 50 percent of women and 40 percent of men are not active enough. Another WHO study says that inactivity accounts for 5 percent of all deaths. That means five out of the next hundred people that you meet are going to die because they weren't active enough.[66]

Inactivity is literally killing us, and it's happening through our phones and TVs and computers. Combat it by making walking part of your everyday life. You don't have to trick your mind into thinking it's exercise. Just get up out of your chair and move around. You'll likely find spiritual benefits to this kind of movement as well—such as the way pain and resistance leads to greater gains and growth. When you exercise, you experience a tiny momentary situation of pain that allows you to breathe better and have more energy later in the day. That's an incredible spiritual lesson to absorb every time you exercise.

Exercise also amplifies positive intentions. People who have felt stagnant, depressed, or uninspired are usually people who don't incorporate regular exercise into their lives. Movement helps you feel better about yourself and have better thoughts, which create a better reality. So take a walk, go swimming, go to the gym, go to yoga, practice martial arts or tai chi, or whatever you need to do. Plan an exercise program and follow through.

Sleep. Cycles of ninety minutes continually appear in our body's rhythms, and sleep is no exception. Could it be that we receive a quantum wave every ninety minutes,

---

[66] World Health Organization. "Physical Inactivity: A Global Public Health Problem." *Global Strategy on Diet, Physical Activity & Health.* Accessed 2019. https://www.who.int/dietphysicalactivity/factsheet_inactivity/en/.

and that's why our energy flows in that pattern? I don't know. What I do know is that the length of sleep is not what causes us to be refreshed upon waking. The real key is the number of complete sleep cycles that we enjoy.

Each sleep cycle contains five distinct phases. For our purposes, suffice it to say that one sleep cycle lasts an average of, you guessed it, ninety minutes—sixty-five in non-REM, twenty in REM, and five more in non-REM. The REM phases are shorter during earlier cycles and longer in later ones. If we were to sleep completely naturally, with no alarm clocks or disturbances, we could normally wake up after a multiple of ninety minutes. You might wake up after four and a half, six, or nine hours, but not usually eight hours.

In the periods between sleep cycles, we are not actually sleeping. It's more like a twilight zone from which we are not disturbed. Someone who sleeps for only four cycles but isn't interrupted will feel more rested than someone who slept for eight or ten hours but could not complete a cycle.

Focusing on and counting the amount of time that you sleep may help you have a better understanding of your energy and how you're sleeping. Complete cycles make you more energetic, give you a longer attention span, reduce stress, and make sleeping part of your overall health routine. If you need to, take a nap. One group of Harvard scientists trained participants to perform a visual task that required them to recognize certain patterns as they quickly flashed across a computer screen. Ten hours later, the subjects who took a ninety-minute nap did much better than those who stayed awake. In fact, they did just as well as people who had a full night's sleep in a previous study.[67]

The Longevity Project. For people who lived long lives, the number one characteristic that the Longevity Project noted was not a genetic trait or something they ate or did. It wasn't because they got over illnesses well, either. Most people who live to an old age do not do so because they have beaten diseases. Rather, they have mostly avoided serious ailments altogether. The best personality predictor of longevity: conscientiousness.

The longest living people shared qualities of prudence, persistence, and organization. You have to care about your life and what's going on around you to live for a long time.

The Longevity Project chronicles these results, which came from 1,500 bright and generally hyperchromic children from 1921 on. The researchers amassed personal information about their histories, activities, beliefs, attitudes, and families. For eight decades, teams of academics maintained the project and continued to assemble exhaustive details about all facets of the subjects' lives. The level of detail involved allowed the study to reach

---

[67] Rob Stein, "A Good Nap, Too, Is Found to Help Retain New Information," *Washington Post*, 2003. https://www.washingtonpost.com/archive/politics/2003/06/30/science/903533c2-ee28-415a-bdec-a5695655526e/.

scientifically sound conclusions—which did not point to cheerfulness or social personability. It was simply that prudent, dependable children live the longest.

Mantra meditation. Transcendental meditation, in which you repeat a phrase or mantra over a fifteen- to thirty-minute period of time, is powerful.[68] According to research, it reduces risk of mortality and stroke in men and women with coronary heart disease. These changes in clinical events were associated with lower blood pressure and less psychological distress.

Color healing. This is a brilliant technique that you can do anywhere, anytime. The idea is that when you meditate, you focus on a color that has some kind of health properties attached to it. If you imagine a ball of red as opposed to a ball of blue, your body responds differently. People have used blue to reduce inflammation, for example. Hospitals are learning that colors have frequencies and can create feelings, and are adapting to use color more intentionally as well. Some people are more sensitive to colors than others. If you are sick, maybe it's the colors in your environment.

Shift your perspective. Look at your body as a system and not just a single organ. If something is going wrong, understand that it may be involved with your nervous system or endocrine system, and that involves a combination of other systems as well. This is not to say that

---

68 You can access mantra meditation on our website at https://advancedsuccessinstitute.com and http://therealityrevolution.com.

you can overcome diseases without a doctor, but perhaps your mind can lead you to the doctor who can help you.

Commit to your health. Health should be part of your overall manifestations. Feel your emotions instead of keeping them buried. Sometimes, the emotions and feelings we have are uncomfortable. That's okay. Have the feeling. Experience it, and let it go. Your health depends on it.

Breathe. This might be the most important thing to come to grips with in terms of fixing our bodies. When you breathe in and out it can completely change your body. It is a constant interaction with the environment around you and should be conscious and focused. Breath can change the way that the body heals, simply by noticing it and not controlling it. Try to become a neutral observer of your breath. Let it become you. As you connect with your breath, as it happens naturally without you, it creates a better connection to your subconscious. If your breath can take you to zero point, you can begin to heal.

The pranayama in yoga uses breath to energize and heal, as well as spinal pranayama which focuses on the intricate nerve complex from your anus to around your pineal gland, focusing on that pathway as you inhale and exhale. The Wim Hof technique uses controlled hyperventilation. Dr. Joe Dispenza's Kundalini breathing technique squeezes the muscles of the body to bring energy to your head.

Then there is the four-by-four method, which is

breathing in for four seconds, holding it for four, breathing out for four, and holding there for four. A long time ago, mystery schools and magic schools trained apprentices on the four-by-four before they would share any secrets. Then there's fire-breathing in and out of your nose quickly in the morning to create more energy than coffee.

When I interviewed him, Dr. Dawson Church explained to me that the best way to create coherence between the neurons of the heart and the neurons of the mind is to begin by slowing your breath to a six-second inhale followed by a six-second exhale.

Each of these breathing techniques allows you to slow your brain and improve focus. Experiment with all of them, and eventually come into a better understanding of breath as a way to understand life. I've had the unfortunate experience of sitting with people who are dying, and when they have nothing else left at all, you can hear their breath. The death rattle indicates that eventually you choose to stop breathing or breath chooses to stop breathing you.

Breath interacts with our thoughts, and by focusing on breathing we can change our thought patterns and begin to break the emotional loops that we all experience.

Listen to your body. What are your energy cycles like? When are you best? What do you need to make your life more productive? How does your body respond to stress, tension, and fear? Notice these things about your body, as a neutral observer trying to gain a better understanding.

Don't be afraid to stretch or breathe or meditate in order to specifically focus on your body.

Love yourself. As you meditate, go through every single part of your body, bones and cells and all, and tell it that you love it. Tell your toes that you love them. Tell your colon that you love it. Stand in front of a mirror naked and tell yourself that you love you. The older we get, the more we cover ourselves up and become ashamed. How can you improve the health of the body that you're ashamed of? Love yourself, no matter your situation, and that love will heal you.

Qigong. We talk about qigong a few times in this book, but I have to mention it here as well. Qigong training has many physical and health dimensions. Robert Peng, a qigong master, was given the choice to develop his skills for combat or for healing. Although he chose healing, he also trained in many techniques to fortify the body, including focusing energy through his fingers to generate electrical currents, or using guardian qi to increase the body's resistance to illness or energy.

Master Peng teaches breathing practices including resistance breathing, which calls to mind the resistance we enjoy when we sing. He also does longevity walking, which is breathing in conjunction with long walks. These are all simple, uncomplicated movements that increase energy and increase the connection between body and mind.

Laugh. One of the greatest things you can do to give

yourself a healthy body is to laugh. People who laugh live longer, are happier, and are more optimistic. For a long time, people believed they needed a reason to laugh before they could. But the benefits of laughter are so great that laughing for its own sake has become increasingly popular. People join laughter clubs, do laughter yoga, and learn laughter exercises. Even fake laughing can get us back to those times when we have healed through beautiful, creative, happy moments.

Create. Creativity puts us in the right state of mind for health. When people are in the middle of a creative project, it's almost like they don't allow themselves to get sick. In fact, if you don't create, you might get sick for lack of an outlet. Maybe there is something you need to express. Sometimes that stuck emotion is a desire to tell a story or display an image. Without an outlet for that creative energy, you might find yourself sick.

The one thing we have total control over in this life is our bodies. Every thought, every movement, every action is a constant manifestation, and to ignore it is to die.

My dad died of Lewy body syndrome, which is a dementia that no one should have to experience. I have wondered if he chose this after my mom died. He didn't lose his memory at first, but he lost executive function. He gave up, and that naturally expressed as this odd dementia. He used to love to exercise, but once he thought he was going to die, he didn't think he needed to exercise anymore. He just gave up.

Please don't give up.

If you are in that moment where you've given up and feel like there is no hope, no reason to hold to the luxury of living, know that everything is going to change soon. It's worth living long enough to see the incredible new world that we're about to experience. Hold on through the reality revolution that is coming. Everything is about to change.

CHAPTER NINE

THE PROSPERITY REVOLUTION

"The desire for riches is really the desire for a richer, fuller, and more abundant life; and that desire is praiseworthy. The man who does not desire to live more abundantly is abnormal, and so the man who does not desire to have money enough to buy all he wants is abnormal."

—WALLACE WATTLES

I suspect that some of you—maybe most of you—will turn to this chapter before reading the rest of the book. Most people are drawn to reality hacking in order to create money and wealth, even if they don't want to admit it. What I've found interesting when coaching people around prosperity is that most of us know very little about ourselves. The way you think about money affects the way you receive money, so if you haven't read the other chapters on blocks to manifestation or reality hacking, I recommend starting there and working your way back to this point.

Money is complicated. It is a living energy that, like it or not, we need to be aware of on an ongoing basis. This is prosperity consciousness, which is more like a relationship than a transaction. Often, if left unaddressed, that relationship becomes abusive.

Amongst the trillions of possible realities that exist, there are several in which you are unimaginably wealthy. You can tune in to one of those realities by using AURA and activating a timeline of wealth. You can then go through the effortless process of actualizing it into physical reality.

The first question you have to ask yourself before moving into a prosperity revolution is this:

Are you willing to be wealthy?

Analyze the way that your body and mind feel when you ask that question. Do you feel any pain, heat, or energy in your body? Do you find yourself itching? Do you find yourself uncomfortable or shifting around? What is your reaction, and what is your response?

The next thing we have to ask is what wealth means to you. How much do you need to be wealthy?

Some people feel wealthy with $5,000, others need $100,000 or a million or a billion. Find a specific number.

If you don't have one, you haven't thought enough about prosperity yet. Before moving on, you need to take the time to think carefully about these things. In fact, it's a matter of sheer survival—it's your obligation to understand your relationship with money. There is no wrong answer, as long as you're specific and honest with yourself.

If you don't understand the differences in money, these numbers won't be accurate or realistic. It might be easy to say you want a million dollars, but I've found that people don't always know what that looks like. Amounts above $100,000 involve a level of finance that becomes quite complicated. You can't treat it like you would a five-dollar bill.

If you're struggling with these questions, try thinking about it in these terms: if I were to give you $1,000 right now and told you to spend it entirely on something that makes you happy, what would you spend it on? You can't save it, it has to be spent on yourself, and it has to be spent all in one day. If you're ready for the prosperity revolution, you would know the answer immediately. Yet most people don't know, or they find it difficult to decide, or they simply have a feeling about what it might be.

If you have your answer, what does your body feel like when I give you that money? How would it feel if I gave you the same amount every day for a week and you had to keep spending it every single day? What about an entire month?

To put that into perspective, it would take two years

and nine months of this exercise to reach a million dollars. To do the same with a billion, it would take 2,739 years.

For many people, this is a challenging exercise. They feel guilt. They don't think it would be right to spend that much. Others can't conceive of the possibility. They won't answer it at all because they are sure it will never happen for them.

Your responses to these questions tell you a great deal about your relationship with money. Are you allowing the possibility of wealth? Can you see where the way you think about wealth may affect the way you receive it or reject it?

Of course, money and finance are completely different, but the same level of open-mindedness and realistic expectations are required. You might be able to manifest a dollar that you find on the street because you've experienced it before and can understand the feeling of it, but very often our ability to manifest larger amounts is blocked because we don't know what it might feel like.

Wayne Dyer wrote in his books about how every day when he would go running or walking, he would find money—every single day. Was it because something magical happened, or was it simply because he had become open-minded to the fact that there was always money around and that he would find it if he looked for it?

What if you could find a penny on the ground and become as excited as if you found a billion dollars? What if that opened your mind to the idea that a force is constantly trying to hand you thousands of dollars with an open hand, and because you haven't had a prosperity consciousness you haven't seen it?

Become willing to accept money. Get in touch with what feelings money brings, and shift them into positive states. This is how we heal our relationship with money to finally achieve a prosperity revolution.

THE STORIES WE TELL

My story with money relates to that majority stat—I have struggled in many ways. I've amassed huge amounts of debt and spent it frivolously on all kinds of ridiculous things. You should have seen my old comic book collection. In retrospect, those things that I accumulated didn't matter as much as my levels of consciousness, understanding, and faith. But at that time, when I had to sell my number one *Iron Man*, *Mad Magazine*, and *Walking Dead* comics, I was devastated.

I had a job that paid well, but I was working long hours and selling bad mortgages. People were locked into negative amortization with increasing balances and prepayment penalties. When I made money, my heart didn't feel right about it. At one point, I held myself back and started to become more honest about what I was doing, and it affected my abundance.

After I listened to a money seminar by Philip Lehrman, who wrote *Prosperity Consciousness*, I began to do internal work, and my income doubled every quarter. I was working more efficiently and creatively. I was reading every book that I could about how to make money, including autobiographies of people who were wealthy. I got into their vibration and energy through their stories. I listened to them talk. The more I reorganized the way I thought about the world, the more I accumulated that energy, and wealth followed.

A great book to read is *MONEY Master the Game* by Tony Robbins, in which he speaks to the reader as if they are already rich. He gives you investment advice by stepping beyond how to accumulate it and into an assumption that you already have wealth, so now what are you going to do about it? This is the mentality of prosperity. It's a game that literally any of us can play—I believe it will revolutionize the world and should be taught as early as school years.

Look back at your past and way ahead into your future. We all have our own stories around money. You might be making money right now, but what is the story of your heart telling you? Money always comes easier and better when your heart agrees to it. Balancing forces are at work, and what's happening internally will come back around to you. Your heart does not know dollar figures, it knows feelings. That is what we are tuning in to.

KEYS TO START WITH

The first step in the reality revolution is to love yourself. We each have an identity of ourselves, and if that identity is not wrapped up in love, it will jeopardize wealth. Stop the desperate, frenzied, hassled chain of chasing after the world's good and instead focus on honoring yourself and treating yourself with care, respect, and approval. When you stop chasing energy, energy starts chasing you.

The second key to prosperity consciousness is to love others. Once you love other people, you'll receive more gifts, money, and attention than you could ever cope with. Of the many keys to being rich, these two will bring you prosperity quickly.

TENETS OF PROSPERITY CONSCIOUSNESS

Imagine someone knocked on the door, and when you opened it up there was just a giant box on the step. It's so heavy that you have to get help bringing it in. Once you drag it inside, you see a note on the top that just says, "This is yours," with your name on it. You open the box and see that it's full of $100 bills wrapped up in bundles, *Breaking Bad* style.

What would your first feeling be? Would it be a fear or suspicion? Dig down to that feeling. There's some part of you that you need to understand before you can alter your consciousness, and it usually involves deep emotions and preconceptions around money and what it means to have it.

On TV and in movies, the rich are identified as evil. Religious courses teach that being wealthy will keep you out of heaven. (Interestingly, "the eye of the needle" was just a gate in Israel where people would walk through with their camels. Most biblical assertions about prosperity have been misinterpreted. Being wealthy is not a sin.)

We can all make plenty of excuses about why money is bad. I've heard people say they wouldn't want to win the lottery because it would be dangerous or they wouldn't know what to do with it. Others talk about how much taxes they would have to pay. In reality, these are easy objections to overcome. Historically, people who make larger amounts of money pay less in taxes because the system can work for you better. You can hire someone to help you spend it and to keep it safe.

Money is not actually the root of all evil. The truth is that money can and will always offer a solution to whatever problem that you have—but when you think back on what you've been told either directly or indirectly, that might be hard to believe.

Not only can money solve any problem, but I believe in a future of full abundance.[69] The discoveries we're making in quantum computing, physics, and technology are moving us exponentially toward a future where we use resources in better ways, where everyone will be rich.

---

69 Once again, I recommend Ray Kurzweil's *Fantastic Voyage* to imagine such an incredible future.

If that future seems impossible to you, consider what that says about your beliefs around poverty. Do you assume, like most people, that someone will always be poor or struggling? This is part of a feedback loop connected with fear, hatred, and anger that we have the ability to change. We will never earn our way to freedom—it has to come from passive forms of income and our acceptance of them.

Once we accept prosperity, that energy moves away from us like a wave covering the whole world. Our belief makes belief easier for everyone else. Changing our thinking flips a switch that can help the whole world.

> **JOURNALING PROMPT**
>
> If you have negative thoughts about large amounts of money, what emotions do you attach to the money that you have? Do you feel worried when you look at your bank account? Do you focus on spending money instead of creating it?
>
> How does it feel when you spend? Are you throwing it away trying to be happy or struggling to survive? Do you give yourself an excuse to spend, such as "It's okay because I worked for this"?
>
> If all you think about is spending rather than creating, with a string of negative emotions attached to it, what's the likelihood

> that you'll have money at all?
>
> How do you feel when you ask for money that is due to you? Are you hesitant to ask for more? Is it difficult to ask at all?

THE ENERGY IN MONEY

We cannot understand prosperity consciousness until we acknowledge the consciousness itself. There are all kinds of emotions and energy around money and the way that we use it. Just think about the way your parents spent—what message were they broadcasting? Did they have fear, hope, or prosperity? Their relationship with money has been imprinted on you, from as far back as the womb. It has given money an electric feeling that you can notice in your body, toward the positive or the negative. For many of us, the feeling is that *it's not okay to be rich*.

I would have so much more in my life if I had not spent years carrying the belief that wealth is wrong. We all would. When I've met people through coaching and at events who have lost or spent all of the money given to them from their family or who had failed investments, they all had a belief that they didn't deserve the money or didn't deserve to be wealthy.

In *The Science of Getting Rich*, Wallace Wattles changed my perspective entirely. This passage in particular explains what we need to undo around the guilt and shame of prosperity:

There are three motives for which we live. We live for the body, we live for the mind, and we live for the soul...We see that real life means a complete expression of all that a person can give forth to body, mind, and soul. Whatever he can say, no one can be really happy or satisfied unless his body is living fully in every function and unless the same is true of his mind and his soul...A person cannot live fully in body without good food, comfortable clothing, and warm shelter, and without freedom from excessive toil. Rest and recreation are necessary to physical life...

To live fully in soul, a person must have love, and love is denied fullest expression by poverty. A person's highest happiness is found in the bestowal of benefits on those he loves. Love finds its most natural and spontaneous expression in giving. The individual who has nothing to give cannot fill in his place as a spouse, or parent, or citizen, or human being...It is therefore of supreme importance to him that he should be rich. It is perfectly right that you should desire to be rich. If you are a normal man or woman, you cannot help doing so. It is perfectly right. You should give your best attention to the Science of Getting Rich, for it is the most necessary of all studies. If you neglect this study, you are derelict in your duty to yourself, to the world around you. You can render this world no greater service than to make of yourself.[70]

---

[70] Wallace Wattles, *The Science of Getting Rich*. (Penguin, 2007). [paraphrased slightly]

Pay attention to how you feel the next time you pull out your wallet or checkbook. If you're like 77 percent of all Americans, you probably feel worry and struggle. All of these emotions are constantly projected outward, even when they seem internal. Your subconscious mind sends them out, and those frequencies are attracted to others that are similar. Subconsciously projecting energy that says "I'm not quite ready," could block you from receiving the money that you want.

THE LAW OF ABUNDANCE AND PRINCIPLE OF INCREASE

The universe is abundance. We have trillions of galaxies with trillions of planets, each one constantly growing and flourishing. Plants grow, animals repopulate, and everything seeks expansion. It is always increasing and growing—and you are part of that universe. That's what life is. It is doing more, being more, and having more. Put another way, God is a class-A hoarder. Every single black hole has another universe behind it, with parallels of realities surrounding us. Nothing is wasted.

In order to develop, things must increase. If you're interested in attaining happiness, constantly increase happiness. Look for new and greater levels, not as a straight line but as a constant wavelike motion with an overriding pattern of increase. This is prosperity consciousness, and arguing that it's okay to be without money is to argue in favor of the stress and unhappiness that comes with lack rather than the growth and joy that comes with increase. We have to change our mindsets entirely. It's okay to be abundant. To be rich. To explore infinity, because infinity is where we understand true wealth.

The Prosperity Bible says, "In nature is abundance for all. Poverty is no part of nature's plan, but the very reverse is true. Nature designed abundance for all. The provision for a man's wants covers not only his necessity, but his super abundance is a law of nature's beneficence."

This is a generous world with more than enough resources for us all. Resources flow in a never-ending stream, matching whatever we think about and focus on. As we develop prosperity consciousness, constantly remind yourself of that: you live in a reality of incredible abundance, you deserve to be wealthy, and you can use that power to attract wealth and plenty, or debt and poverty.

RECIPROCITY AND VACUUMS

You are an incredibly unique person in this universe. No one else has the genetic combination and life experience that you have. You are powerful and deserve to be rewarded for it. You deserve to be wealthy.

If it feels uncomfortable to say those things, remember also that you cannot give to others without first having wealth to give. When we identify others' needs and share with them, we can all become abundant. This is the law of reciprocity.

Connected to reciprocity is the idea of vacuums. Just like we want to help people without obtaining wealth ourselves, we tend to meditate about new furniture while sitting on the old couch. Creating an open space leaves room for the vacuum to be filled.

If you want to have new clothes, give away what's in your closet and watch how it's filled again. If you want new furniture, get rid of what you have and watch what happens to the open space in your house. For wealth to work, it has to circulate. Using and giving the money and things that you have creates a vacuum for that circulation to continue.

Imagine knowing with certainty that every time you give something, it will come back three-fold. This is reciprocity and vacuums in action, and it's a key component to prosperity consciousness.

THE TRUE MEANING OF PROSPERITY

Deepak Chopra said that "life is a field of unlimited possibilities," and we have to be open to that being true in order to engage in the prosperity revolution. When we are aware of limitless possibility, we become more interested and curious about things that we otherwise would not notice. Just like wanting to find a twenty-dollar bill on the street will help us see it, we have to look for infinity to find it.

We're discussing prosperity in terms of monetary wealth, but more deeply, it's a state of consciousness. Steps to wealth creation are popular—scroll through a social media feed and every sixth or seventh post will be a money-making scheme. But "do this and you'll become a millionaire" is backwards thinking. The idea that results are the product of specific actions taken over time is only half the truth. Real success takes one more step back to ask where those actions stemmed from in the first place.

Why haven't you lost weight, started the business, traveled the world, or bought the house that you have wanted for so long? It's not because you don't know the right actions to take. It's because of your state of consciousness. The right actions come naturally when your consciousness is one of prosperity, not poverty.

Self-made limitations—excuses—prevent us from doing what is in our best interest. Prosperous minds don't make excuses.

Stepping out of a monetary context, if you want to lose fifty pounds but still think of yourself as someone who is fifty pounds overweight, that's who you'll be. As with every other concept in this book, you have to both feel and see the outcome you want in order to make it happen. Not a vision of the future, but a feeling that the outcome is happening and real right now.

This is how we operate from an abundant state of consciousness rather than one of lack, and it is the key to unlimited financial success. When you reach that sweet spot where you know in your heart something is going to happen, belief begins to take over. All of humanity's creations have begun in this way, in the mind of one individual who believed it could become reality. Only after clarity turns to belief can actions be taken.

THE FOUR STAGES OF PROSPERITY

Money is the ultimate example of particle wave duality, which we explained in chapter 1. When we're thinking about the money we want to make, what we have to do to make it, and how we will spend it, that money is in wave form. Once we actually make the money, it collapses into a particle. Moving it usually happens in four stages: unconscious incompetence, conscious incompetence, unconscious competence, and conscious competence.

First, we don't know what we don't know.

Second, we discover something brand new that we didn't know existed, we're now aware of that potential even though we don't know what to do with it.

Third, we can start practicing it, attempting to be competent though we don't know how to navigate it with any amount of expertise.

The final stage is mastery. If you drive a car regularly, you don't have to think about how to do it anymore. You experience the drive, but by the time you reach your destination you hardly remember the journey. You're unconsciously competent.

Prosperity at any level usually happens through unconscious competence. It's the result of beliefs and repeated actions so thoroughly and for so long that it becomes unconscious. Happiness is earned by practicing happiness, not by waiting for a stroke of good luck. If you're not happy, you likely have habits of thinking, feeling, or behavior that make you that way. The same is true for prosperity. If you're not where you want to be in life, it's because of your habits. We have to create better habits and beliefs and apply them to our lives every day—a conscious effort until it becomes an unconscious part of our lives.

THE IMPORTANCE OF CONTENTMENT

There is a point of balance between wanting to be rich and being strongly dependent on money. Intense desire comes from a level of lack, which is counterproductive. Accepting what you already have and realizing things can always be worse is a good place to begin. You don't have to stop wanting money entirely, but you should relax about the fact that it isn't pouring in like a geyser right now. Take the position of the gambler at the table who could easily hit the jackpot or lose everything, but who is fine in this moment.

Do not postpone positive feelings for the future, when you might possibly have everything you desire. You do not achieve happiness by manifesting your desires—you manifest your desires by achieving happiness. As within, so without. There is no reason to wait for any time in the future. That future never comes if all you do is wait for it. Feel prosperous now. Feel confident now. Feel happy now. Make this day a joyous celebration of your inevitable success.

Envy can keep us out of the moment, and it is one of the worst feelings you can entertain. If someone has more money than you or has a similar job but makes more, remember this rule: if you're envious of anyone for their wealth, you can never have what they have. The moment that you say that what they have is bad, your subconscious hears that it is bad and will work to make sure you never get it. If you find yourself transmitting envy or jealousy, reverse it as quickly as possible.

Competition is not necessary in an abundant world. You are here to create, not to compete for what has already been created. You are going to get what you want, but in such a way that it helps everyone else as well. You don't have to take anything away from anyone else, and they aren't taking from you. You don't have to drive bargains, cheat, or take advantage of people. No one should work for less than their value, and no one has anything that you cannot have yourself.

This kind of radical contentment is a difficult emotion to hold. It's hard to love money without wanting more of it, and it becomes almost impossible to avoid a dependent relationship. Vadim Zeland says, "All one can do is attempt to limit the aspect of dependency to a minimum. Be happy if money has come to you, but never kill yourself worrying about not having enough money, or spending it, otherwise, you will have less and less of it."[71]

If someone doesn't earn much money, they probably also make the mistake of complaining that they'll never have enough. The parameters of this type of thought energy often correspond to lifelines in which there is no wealth. It pulls you into a reality that matches your beliefs.

This is particularly dangerous when you live in fear and anxiety about your income gradually decreasing. When a person fears losing money or not earning enough, they activate the most effective method for shifting lifelines, but in a negative way. They begin to actually have less and less money. You bring your attention and focus toward the fear and it pulls you into matching realities.

THE POWER OF THE PENDULUM

I'm aware that there are people making money in opposition to all of these ideals. They are taking, competing, and harming people. If you recall the power of pendulums, know that they use money as a way of rewarding adherence. In the material world, almost anything can be bought and sold, and it becomes a falsely glittering bait

---

[71] Vadim Zeland, *Reality Transurfing: steps 1-5*. (Ves Publishing Group, 2012). Translated by Joanna Dobson.

that can pull us into lifelines that have money, but also have nothing to do with our purpose or true happiness.

When you're overly dependent or expectant about money and it's coming in anyway, question whether it is coming from pendulums, which often leads to the idolization of money on its own. Don't allow money to become an artificial substitute for your actual goals. Pendulums gain power from this dynamic, while the individual often loses their way.

When you're pulled into that timeline, it's easy to forget about the trip or the achievement or how you wanted to help people. It's easy to lose a sense of what you really want from life and to instead focus on the search for money. Wealth should never be a goal in and of itself. What are you trying to take away from it? What experiences do you want? What joyous events do you want to experience?

Your ideal lifeline is impossible to access from someone else's pendulum. They function outside of the context of true purpose. They serve someone else's objective, not your own. Ask yourself where the money is coming from and whose goals it serves. The true prosperity revolution will make money an accompanying attribute, with the real reward centered on what we each want from life—buying a house, traveling the world, taking that camping trip in Alaska or the skiing trip in the Alps, or raising horses on your own farm. Go after the goals that a sack of money can get you, not the sack itself.

AVOIDING POVERTY CONSCIOUSNESS

Without entering into this state of mind, let's think about what it actually entails. A poor person only sees the external side of wealth—the luxurious houses, the expensive cars, the diamonds. A poor person in this environment would feel uncomfortable. If you were to give them a suitcase full of money, they would do all sorts of foolish things to spend it all.

This level of consciousness transmits a frequency of energy that is sharply dissonant against a wealthy life. Until a person can let the attributes of wealth into their comfort zone and to learn how it feels to be the owner of expensive things, not even buried treasure could keep them from being poor.

When you have this poverty consciousness, dramatic events can unfold that induce a transition into terrible lifelines. The most insidious thing about this induced transition to poverty is that the spiral begins to unwind slowly then suddenly picks up speed until there's no stopping it. This is when a successful person starts with some financial difficulties then suddenly loses everything and ends up on the street.

Setbacks will happen. Things will affect you in negative ways. If you do not get angry, depressed, excessively anxious, or resentful or deprived, then you will not give that spiral the energy it needs to gain speed. The pendulum that's affecting you, or the wave that has picked you up, will come to rest. An induced shift into poverty consciousness can only occur when fear takes hold. It can only move you if you react to the destruction in the moment.

> To check in with yourself, think about how you act when you receive a large amount of money. Do you act differently? Do you have more ego? Are you brash? These are signals that your relationship with money is out of balance.

UTILIZE AFFIRMATIONS

Creating an affirmation is creating a thought. Use your affirmations to direct your focus to beauty, even when the tasks are mundane. Use them to shape your prosperity consciousness and project the reality that you want to manifest. Recording them in your own voice can carry a lot of power as well.[72]

Acknowledge that you desire a deep consciousness of financial freedom. You desire the flow of prosperity to become equalized, and you desire a greater ability to attract money and to find success in whatever you're doing.

You can afford to use money for happiness. You have no question of expenditure of clothes, food, books, and entertainment. You have what you need for health, happiness, friendship, and service. You are willing to be uncomfortable now so that you can be extraordinary later.

---

[72] *The Reality Revolution Podcast* has several episodes with affirmations to use.

You aren't chasing money, but are allowing yourself to have it.

As you hone your affirmations, make sure they are specific and realistic. If it doesn't feel realistic to you, you will project the opposite energy, so keep your feelings about the phrasing in mind as you say them. For example, if you start with "I'm a millionaire," rephrase it to, "My yearly income doubles." Be so specific that you overcome anything that might work against you. For example, someone might say, "I have money coming to me regularly," and wind up with small or large amounts of money coming in the form of debt.

I created three hundred affirmations on wealth and recorded them in my voice.[73] If you aren't ready for that, there are several compilations that you can access on YouTube for free. I listen to them on three-times speed during my workout. If I do that regularly, they come together in my memory in ways that shape my thinking.

Over time and repetition, you will begin to believe that money is good, that it makes you happy, that you deserve it and can handle it, and that it is possible for you to have whatever you want.

TOOLS FOR A PROSPERITY REVOLUTION

Choose abundance. Thoughts are things, and powerful things at that. One of the most common causes of failure,

---

73 Some of my favorites are included in the back of the book.

especially around prosperity, is the habit of quitting when you've been overtaken by temporary defeat. When those moments tempt you to spiral in defeat, start to notice the abundance around you.

When you get in the car, look at the beauty around you. Notice the sun, water, and grass in abundance. Ponder the infinite. It changes everything.

Write a letter to the universe. Tell the universe everything you want, without shame. Create the world that you want, and make it specific. Put that letter somewhere that you can't access it regularly, then come back to it in three to six months. You'll be amazed at what communication with the universe can accomplish.

Respect the law of flow. If you hold your money in, it will never come back to you. It's okay to spend without feeling guilty. Something that I like to do in a group or a seminar is to take a ten-dollar bill out and have someone offer their services for that cash. Then they do the same, and you watch as it flows around the room. If you could watch the money that you spend flow, you could see the propulsive power that it has, and you could trust that eventually it will come back to you.

So whatever you want to do, go for it without any guilt or hesitation. Create the wealth that will afford the life you want to live. Make the money with your heart, consistently, being open to the new ideas that will inevitably come to you. When you open yourself up to the idea that you can make more than what you have, that more is on the way, you can let go and begin to enjoy the flow of prosperity.

Get uncomfortable. If I were to choose a common trait in the people I know who have become rich, it's that they do the things that make them uncomfortable. Lean in to hard things for short periods of time.

Accept delayed gratification. If I were to offer you $5 now or $15 if you wait three months, researchers have found that many people will take the instant gratification. Prosperity consciousness is willing to wait. When you think you need it now, it transmits the power of lack. Allow yourself to be wealthy, even if it takes time to realize that state.

Learn to receive. When you go to a restaurant with someone and they offer to pay for dinner, how do you respond? When someone offers you a gift, what do you say? A lot of times, we block what we want in life because we're simply not willing to receive. If someone offers you something, take it thankfully and gratefully. Don't cause a fuss. You deserve it—open yourself up to live in a receptive mode.

Open your eyes. Start looking for the money that's already there. You've been given billion-dollar ideas every day—and maybe you've already seen them come to fruition through other people. Be ready to be the person who carries them out.

One reason that the 99 percent are not rich is that 99 percent of us don't pay enough attention to detail. Poverty consciousness doesn't have the energy for detail. It doesn't want to read the contract in full or study where the money is flowing or break down things that feel complicated. Do you know what your bills are? Do you know what your monthly expenditures are? Do you know exactly what you're making?

All of the means of the universe will come together for you if you keep your eyes open and follow opportunity. Focus on the wave of prosperity, without a fear of failure, and you will get better and better at it over time. It's your duty to get rich—don't miss opportunities because your eyes and heart are closed to them.

Open five accounts.[74] In one bank account, put all of the income you receive from any source. Then create another for regular purchases and daily debts, another for large purchases, another for investments, and one that you'll never spend—leave the last one for your kids and your financial legacy. If you only have a penny for each account, get them all open and functional. Then work up to a point where you can spread 10 percent of your income across each of these accounts.

When you see the large purchase account starting to grow, you'll start thinking of things that you want to use it for. It allows you to delay gratification and start planning and thinking in larger terms.

Remove debt from conversations. Conversations about money always seem to turn toward debt. Everyone wants

---

[74] This technique comes from Lehrman's *Prosperity Consciousness*, which I highly recommend.

to know how to deal with it, but if you focus on the debt itself, it will accumulate. Try to remove debt from your conversation and instead think about your income increasing. Find ways to automate your debt payoff so that you're not turning to it so often and broadcasting "debt" as a frequency. Focus on how excited you are when the bills are paid off rather than the debt itself.

Channel money. I believe that money is not only an energy but a conscious energy—a living spirit that can be channeled, just like Esther Hicks channels Abraham. This is one of my favorite exercises. Sit down with a journal and a clear mind. At the top of the journal, write down, "I want to have a discussion with money."

Then hold your pen at the paper.

After five minutes of silent meditation about what money might want to tell you, start writing whatever comes to mind. Don't worry about what it says, just try to get into a conversation. Money is a very old energy, and it wants to love you—but it will not respond to you if you really, really want it. It's the girlfriend who wants to play hard to get. If you really want it, you can't have it, but if you're open to it and create an intention, it will be there.

Even if you won't believe in its consciousness, do the exercise. Ask money questions as if it could answer. If you tell money it's evil, it won't want to hang around you. Tell money that you love it just as you would tell a person—not to abuse it or use it, but to appreciate it.

Sexual transmutation. In *Think and Grow Rich*, sexual

energy is included as a part of prosperity. There are ten different things in that book that can stimulate the mind that have led to prosperity for other people, and the first on the list is the desire for sexual expression. Further on the list there is love, friendship between others of the same or opposite sex, and mastermind alliances. He gives examples of people with high sexual energy who are better at sales, because there is something to that energy.

The open, happy joy that we feel in sex is similar to what we feel when we receive large sums of money. To bring that energy into your prosperity intentions, you don't have to have sex. You might create a playlist of music that you enjoy during sexual activity that you also play when you're making money decisions. Anchor some of that energy and bring it into prosperity.

Find inspiration. Every day, spend fifteen minutes taking in something inspirational. Just like you shower and brush your teeth every day, don't miss this either. Look at something inspirational that can cleanse your mind. Listen to inspirational books and radio programs about financial health. Stop listening to the news and replace it with topics of success and education. Access and accumulate energy on a regular basis that can counter all the stuff we're hearing about how hard it is to make money and get out of debt. Constantly expose yourself to prosperity with books, messages, and affirmations.

Daydream. We have no problem engaging in negative visualizations and scary thoughts, which could be why 95 percent of us are not truly happy with our circumstances. Instead, use that energy to set aside time to daydream about your own financial health. This is actually an intelligent process, not childish. Don't listen to anyone who says otherwise. Imagination is fuel. Whatever you imagine, you feel, strengthen, and make more likely to manifest in your life. Spend as much time as humanly possible imagining the outcomes you wish to achieve.

If you're not imagining what you want, then you're probably achieving the opposite. Plan out your perfect life. Write it down on paper as if you've already achieved it. Write in present tense, capturing the daydream and clarifying it with details and specifics.

Check your words. Start writing down everything you say to yourself or others about money. Then reverse what you have said that's negative. If you say, "I can't afford it," notice the power behind those words and switch it to something more positive. Money hears you saying what you can't do or how much you shouldn't have spent or lost, and responds.

Enjoy the journey. Sometimes we create goals that take over the journey. They get bigger and bigger until we forget that this is about the experience not the accomplishment. Achieving the goal is not all there is to prosperity consciousness, and if we lose sight of that, we lose the consciousness that we need.

Redefine wealth. In my twenties, someone who was extremely wealthy told me that I didn't have any money because I thought about it all the time. Ironically, the only thing that stands in the way of money is wanting it. Instead of thinking about money, think about things that are important.

To be wealthy means that you have a large amount of something with which you do work that can benefit others. Masaharu Taniguchi said, "What works to benefit others is love. When we put love into practice, it becomes wealth." Once you let go of wanting large amounts of money and instead pursue true wealth, you'll find it.

Focus on already having. Before something happens in the "outside world," it has to be claimed as reality within. Before you can move toward that parallel reality, your belief has to guide you like a GPS toward what you want, as though you already have it. If you struggle to identify what it feels like to already have something, try this.

Sit back and close your eyes. Softly focus on something you already have or own. Now let something come up that you like or appreciate, then something else that you love. Remember something that you appreciate. Recall something you're fascinated by. As you think of these things, notice the separation between the object and the desire. Merge with what comes to mind and allow yourself to feel what it's like to already have those things. Relax into that sensation of having and being grateful.

Defuse money. In another meditation, try to view items neutrally, without label and without resistance. Allow images of money to arise as well, but do nothing with those images at all. Look at the thought of money until you feel no more charge. Keep going until you feel no more urge to quit, no desire to change it or get away. Stay with it until it becomes neutral. Then intentionally recall a time someone gave you money, a time you gave someone money, a time someone gave money to someone else, a time you withheld money from someone, a time someone withheld money from you, a time someone stole or a time that you stole. Recall a time in which poverty solved your problems, a time when it was comfortable. A time when money solved problems and one when it was comfortable. Think about other people having money and power that you do not. Think about the problems you'd have with a lot of money and power. Think about granting poverty to others and granting money to others.

Observe the lies, suppressions, abuse, greed, blackmail, and betrayal around money until you can observe it all with neutrality. This exercise will defuse the emotions and stuck beliefs surrounding the subject of money—neutrality opens you up to focus on what is prosperous.

Dissolve negative belief patterns. Acknowledge the thoughts that have condensed into beliefs. Say out loud, "I'm afraid of X because Y," or "My life is not perfect because..." or "I'm not rich because..." or "Money will be bad because..." Make a list of all the things you have an aversion against regarding money. Welcome any feelings or thoughts that emerge. Observe the negative thoughts and break their momentum.

Know and reframe each of those into something you can appreciate. Handle the chaos of your mind. The habits you have around your thoughts, beliefs, and actions are all part of your subconscious. The principles we hold on this level will lead to poverty or prosperity. Turn them into the right habits.

PROSPERITY AFFIRMATIONS

Here are some sample affirmations that I have found to be powerful:

- My job is a pipeline by which I tap the infinite wealth of the world economy for my own personal desires.
- I deserve to be wealthy.
- My income increases every day, whether I'm working, sleeping, or playing.
- A part of all I earn is mine to keep.
- My income now exceeds my expenses.
- Every dollar I spend comes back to me multiplied.
- I am released from my self-image to experience infinite potential unfolding from within me.
- A part of all I spend goes into permanent capital, expenses, or reserves.
- I am responsible for everything that happens to me and my money.
- I am financially free—or, I am becoming financially free, or I am allowing myself to be financially free.

- I am wealthy, both spiritually and materially.
- My inner compass guides me to wealth and abundance.
- My wealth is abundant and appears when I need it.
- My mind is a center of divine operation, and I receive all the money that I desire through my spoken word. All doors of abundance and wealth swing open for me now.
- Miracles never cease.
- My life is overflowing with abundance, prosperity, and money.
- My spoken word releases all of the abundance that is mine by divine right.
- Unexpected doors open for me and release all the money and opportunity I need.
- God is my infinite supply.
- Large sums of money come to me easily and quickly, in increasing quantities, from multiple sources, on a continuous basis, in the best interests of all, in a manner that I get to keep it.
- My affirmations are effective, whether I believe them or not.

PARTING THOUGHTS ON PROSPERITY

Your thought process is the key to reaching prosperity in your life. By entertaining thoughts that are prosperous, taking actions that are prosperous, you can change everything.

I cannot end this chapter, however, without acknowledging the question of deep poverty that we see in tent cities and impoverished nations. Can they just think their way into money?

It's going to be much harder for anyone surrounded by poverty. When you breathe poverty in from the time you wake up to the time you sleep, it can be difficult to overcome.

I'm writing this book because I want you to find a way to escape any situation that you're in. It's going to be harder for some than for others. But it's worth it. You're not deluding yourself to believe you are wealthy. It's just the process that you're going through. The beautiful part of it is that your energy will affect others around you, traveling like a virus. You don't have to focus on the poverty or struggle in your life—you can shift your thinking, and eventually your reality will shift into one of abundance.

Don't give up. Don't feel stuck. Don't lose hope. Choose love and prosperity and good health will follow. We have to uplift one another and bring each other up out of these mindsets of despair and lack. You can create the revolution instead of playing into the pendulum that needs you to struggle.

The more you love life as it is, the more you love what you do, the more you will ascend spiritually. Even if you're in the mundane process of making money, look for the beauty. Find wonder, knowledge, and wisdom in the mundane. Hold a mindset of joy in the little things as you work long hours or live a normal life, where even the commonplace is part of infinity.

Everything is a miracle. Love money. Smile at it. Collect it. Enjoy it. Feel its power. Spend it. Invest it. Give it away. Receive it. It is energy, and it is powerful and beautiful and connected in this intricate web of reality.

Stop thinking that there is only so much to go around. As the Reality Revolution spreads across the planet, as consciousness expands, so will our prosperity mindset. The whole planet will begin to come from abundance instead of lack as it realizes the abundance of infinity itself. Infinite realities are available to everyone. Everyone will be prosperous. Everyone is prosperous now. A reality exists and will exist in which everyone is wealthy beyond their wildest imagination. The energy of this book is designed to pull you into that reality. Tune in to these words and feelings and let them flow through you. These are truths that you already know deep down. Welcome to the prosperity revolution.

CHAPTER TEN

THE LOVE REVOLUTION

> "You, yourself, as much as anybody in the entire universe, deserve love and affection."
>
> —BUDDHA

If you can find true love, it's worth more than the richest man in the world can pay. There's nothing more transformative, incredible, and powerful than love. It's a feeling. It's a creative force. It's an energy. It's stronger and more subtle than any of us can imagine. I believe it's source itself.

The biological, psychological, emotional, and spiritual power of love can catch you up in an exhilarating, consuming wave that cannot be replaced by any other experience. You can think of nothing but intertwining with that person, body and soul, again and again. Nothing can possibly be better than love, and many people spend their lives striving to find it.

There is a parallel universe in which you experience the deepest most profound love possible. By using the AURA technique, we can activate and actualize a reality that already exists, in which you find the perfect love that you long for.

Interestingly enough, when you step into the reality revolution community, 90 percent of the people posting about the Law of Attraction and reality creation are looking for a significant other. They even have lingo: *I'm trying to manifest my SP*—their *specific person.*

This poses an interesting conundrum, as we'll explore in this chapter. Is it possible to manifest a specific person? Is it appropriate?

Regardless of the answers to these questions, the desire remains. As soon as people understand the enormous power they have in creating reality through their thoughts, love is one of the first things we want to find.

We live in the greatest time to find love. Never in the history of mankind has it been easier to find the ones we are truly meant for, from any place in the world. We have a global marketplace of people and ideas. There has been an explosion in apps to connect people to each other. Understanding and psychology have made it so we can interact with each other better. Finding the perfect person is so much easier now than ever before.

It's also never been a better time to be single. And

in many ways, being single can be just as wonderful as being in a relationship. The idea that we need to create a reality with a partner to have love is not necessarily true. Love is much more than that.

Love itself is a spiritual lesson of life. We carry so many stories around it, evidenced by the millions of romance novels lining the shelves of bookstores around the world. Each of those stories is about love, as much as this chapter is about love, or the stories we tell ourselves are about love.

I personally believe that we are multidimensional and infinite, with tons of different soulmates. I asked Frederic Dodson about this, and he said, "we have an infinite number of soul friends and perhaps one soulmate. The idea feels good." I truly believe that we cannot get into a lack mentality when thinking about love, as if there were only one person out there that we somehow need to access. It can't be true. Once you raise your vibrations to the point of understanding oneness and infinity, you realize that everyone and everything can be loved. That's where true soulmates are made. Perhaps we are all soulmates.

DATING IN THE NEW WORLD

Apps like Tinder or Plenty of Fish or Match.com are incredible feats of technology. When I entered the dating world again after a long-term relationship, I thought about meeting people in person at bars, clubs, and grocery stores like we did years ago. We had to meet people through friends or intentional connections—we didn't have screens and filters and so many people to sort through. Now, with some intention around the kind of person you're looking for, it's amazing whom you will find.

I had to find a way to date again in this new and unfamiliar world. I approached it the only way I knew how: like a businessperson. Looking at the amount of time it took to use each of those apps, I realized it was the same as any other work that I've done. You have to make the contacts, tell the jokes, get through the initial dating script, then try to meet. It's mind-numbing, annoying, and soul-sucking. It can take you into a different version of yourself. So I began to treat dating the same way I would as if I were manifesting something for my company.

We make it that complex. Dating is at that level of programming and technical thinking, where it's more about the quantification than the connection. We are playing a game where the other person is an object. I'm not alone in this—Neil Strauss wrote a book called *The Game* that I believe everyone should read. It digs into the way men approach dating and breaks it down into a science.

I believe we're filtering out our ability to love—as evidenced by 67 percent of marriages since 1985 ending in divorce. Of the last three weddings that you went to, two of them will end in divorce. For a man especially, that's an expensive prospect. Marriage is a risk and, by association, so is love.

The young adult fiction book *Delirium* by Laura Oliver took our fear of love to its furthest reaches: a world where they surgically removed love from every single person at a certain age. They decided that love was the reason for all of the problems in the world—murders, wars, famine—and that removing it was the cure-all. That's not the world that I want to live in.

Love is more than the overwhelming high of a chance, perfect connection. Feelings will wane and settle into the mundane predictability of everyday life. Natural conflicts and resentments build up over time, and the power that we feel with new love starts to dissipate. Don't think that this burnout is love's conclusion.

The energy in the human body, at its purest potential, has been calculated by physicists to be ten times greater than the energy in a hydrogen bomb, thanks to Einstein's famous $E=mc^2$ using the mass of an average human body. If you're willing to use that energy in your relationship, there's limitless potential to expand beyond the day to day, to bring the power of love into our lives in a way that far exceeds the novelty of the honeymoon phase.

We don't need to remove love or commodify it. We need to experience it on a deeper level.

FRAILING: A COUNTERINTUITIVE WAY TO PURSUE RELATIONSHIPS

One of the most fascinating concepts discussed in

Zeland's *Reality Transurfing* is "frailing" because of how effective it is in finding and maintaining relationships. It works every time. It mirrors some NLP concepts, but in a much simpler manner: abandon the intention of receiving and create an intention of giving, and you will receive the very thing that you gave up.

Understand that most people are only thinking of themselves. What is their self-interest? How can you help them? The feeling of self-worth lies at the core of inner intention. Switch your attention from yourself to others. Try to increase other people's self-worth. Make it a game. By showing an interest in those around you, you will attract attention to yourself. In a conversation, people are not going to judge how interesting you are. They are evaluating how you might suit the role of realizing their self-worth. Be sincere in your interest. Don't ever argue, even when they are totally wrong. Ultimately, by frailing, you use other people's inner attention to achieve your goals.

Everyone has a special signature frequency that you can tune in to that will make it easier to vibrate into a connected state, resonate with their soul frequency by understanding their dreams and their self-worth. You cannot tune in to this frequency by thinking about what it is you want or need. Instead, help increase their self-worth and fulfill their inner intention, and your relationships will change dramatically. Just try it. Stop thinking about yourself and think about those people you want to have a relationship with. Focus on them with the same passion you focus on yourself. This may seem unusual at first, but once you start to see the immediate results, you will undergo a paradigm shift in how you relate to every person you meet.[75]

WHAT KEEPS US FROM LOVE

So many people will tell me that they "can't find love," then go on to attribute it to negative qualities about themselves. They're too ugly or too fat. They always have some excuse. We all know at least one person like this—the one friend has created an image of themselves and uses it to explain why they don't have the relationships or things that they want.

Of course, these excuses aren't true. There are "ugly" or "fat" people with beautiful people all the time.

If our surface excuses aren't true, we need to identify what really keeps us from love so that those blocks can be removed.

IT STARTS WITH YOU

If you're struggling to find love or want to create a relationship but haven't been able to make one work, see if you can first find a way to love yourself. It's better to put

---

[75] For a better understanding of frailing, check out this episode of *The Reality Revolution Podcast*: http://www.therealityrevolution.com/understanding-frailing-in-transurfing-ep-124/specific.

an image of yourself to admire as part of visualization techniques instead of someone else's image to manifest. This is the most cliché yet fundamental requirement to maneuvering into love: if you do not love yourself, you will never feel worthy of love from others.

Unlike other aspects of life, this is not an area where you can simply mimic someone else. You might find a role model who is good at meeting people or starting relationships and then learn from them as a demonstration, but not to emulate them or create a template for your own life. Rather than striving to become just like the person you admire, strive to become yourself. Allow yourself the luxury of this life that you're shaping.

You have a right to your own individuality—a life where you don't simply copy the experience of someone else. Those who try to use techniques to become someone else entirely will never find contentment. Your only measurement is your own soul, not anyone else or their timeline. Wearing the mask of another will at best make you a copy and at worst make you a parody.

The greatest actors did not become great by copying someone else. They are great because they have found themselves. Despite differing characters, all great actors are able to play themselves. It's easier to put on a mask than it is to pull it off and stand in your own identity. Once you wake up from the scripts that are controlling your life, the next question is whether you can exist without the "camera" rolling and scripts telling you who you are. Attempts to repeat other people's scripts are futile. Acknowledge the brilliance of your own individuality, and other people will have no option but to agree with you.

Allow yourself to be presumptuous enough to *have*. Loving yourself in this way invariably leads to self-approval. No one can compete with your unique qualities—you are full of beauty and worthy of love. Give yourself permission to be that person. You might start by repeating affirmations or by looking in the mirror and creating moments where you feel love for yourself.

Choose the love that you want, and accept it. True love is yours for the taking—but it is a state of mind you must move into within yourself first.

EMBRACING YOUR FAULTS

When we try to hide our shortcomings, it keeps us from loving fully. We work to hide that scar or that feature or defect, whether it's visible to others or something that only we notice, so much that we're no longer authentic or genuine.

In chapter 2, I told you about the underbite that affected my self-image so deeply that I chose surgery to resolve it. What I didn't tell you was how extensive the underbite and subsequent surgery was. I never smiled in pictures, and hardly at all for anyone. When I closed my mouth, I could put three fingers from my top teeth to the end of my mouth. I hardly ever opened my mouth in public. I had relationships in that time, but I always held back thinking I wasn't worthy.

While waiting for my body to stop growing so that the surgery would be effective, I had to have braces that made it worse before it got better. When it was time for the surgery, they broke my jaw, moved it, transplanted bone from my hip, then wired it shut for three months.

The first time that I smiled once my teeth were fixed, the confidence I felt told me something powerful about myself: the surgery hadn't given me anything. My natural energy and confidence had nothing to do with what I looked like and everything to do with what I felt. Looking back, I missed out on so many moments where girls that I liked were clearly interested in me but I wouldn't approach them because of how I felt about myself.

Your shortcomings, no matter what they are, will be beautiful to someone else if they are beautiful to you. The false images that we hold of ourselves affect our ability to love, especially when we use them to hide. You've read far enough into this book that you believe you have the power to create your reality—why would you let some perceived inadequacy hold back the power of the universe?

CONTENTMENT WITHIN RELATIONSHIPS

If you're in a relationship at the moment, find a way to love that state. You don't need your partner to do anything, change anything, or show up in a different way.

They don't even need to know that you are doing this, and they certainly don't need to buy in to these concepts. But when you change, your partner will as well.

When you harness the power of your own energy and consciously use it to create true love, you get everything that you're looking for in your relationship. You find the love in your life. You unlock a more evolved version of yourself—one who can handle every obstacle with strength and grace, who lives authentically and freely, and who loves and is loved unconditionally and passionately.

If you can find love in your relationship, you'll start to attract love from others. Be in love. Go through that process. Be open to the love that's coming to you, knowing that fear would only push it away.

CONTENTMENT WITH BREAKUPS

In his book *Conversations with God*, Neale Donald Walsch tells the story of two people, Tom and Mary, who saw each other from opposite sides of a room. As they each radiated their personal energy, the fields met in the middle of the room and began to create a combined vibrational frequency—a relationship energy. The new relationship field gained its own shared purpose and voice. They became entangled. He called it Tom-Mary.

As they both fed energy into Tom-Mary, the energy was sent back to each of them through a quantum field.

The closer they drew to each other, the shorter and more intense the cord of energy between them became. Each step toward each other made the vibrations burn wider, brighter, and deeper. Their own individual vibrations amplified Tom-Mary.

When we have a frequency match with someone that becomes this intense, we get sucked into that field. If Tom or Mary's frequency were to change later, that energy field might not be the same. But they're sucked in. They don't know how to escape—and really, there isn't a way to just leave. You have to find a way to intentionally cut the energy.[76]

When we end up in relationship with someone else, we become entangled. Even if we discover we don't really want to be with someone, we get stuck. It becomes difficult to move on when we share energy in this way. This entanglement can last for years, long after the breakup happened. Sometimes I find that people trying to find love aren't able to because they are still connected with someone in their past. They say it takes half of the relationship's length to recover—I hope that's not true, and that we can advance ourselves to a faster recovery once we understand what we're moving on from.

It is okay to move on from terribly unhappy relationships. The fear of the pain of breakup and the unknown of finding love again keeps us locked in when we need

---

[76] There are meditations at the Advanced Success Institute that might help, including the Soulmate Meditation and Letting Go of Your Other.

to move on. If there's a voice telling you that true love is somewhere else, it's okay to listen to it. Speaking as someone whose partner left for true love, it is painful at first for the person you are with. But once I embraced the idea that there was a love that was meant for her, and that she found and chose it for herself, it became easier to deal with. If your relationship isn't working out, don't make it personal. Someone else's love is not an insult to you.

Let yourself love. Let other people love. We each deserve to create our best reality, and we deserve that reality.

JOURNAL PROMPT

Look back on your love story from a different perspective. How do you understand love and show it, and how did your parents understand it and show it? Is there something creating love in your life that you don't want? What does the story of your love tell you?

THE POWER OF BALANCE AND IMBALANCE

We talked about importance in chapter 2, and it's especially true in the case of love. If you make a specific person incredibly important, or if you make loving them important, you'll usually create the opposite effect that you want. You might have already seen in your life that when you want someone or something really badly, you don't end up with them.

The universe tries to balance things out, and placing too much importance on a partner or the desire for a partner can create opposing forces. Similarly, dependency creates excess potential. If you become dependent on your partner, the need for balance creates problems. If you believe you possess that person in any way, shape, or form, that love will be imbalanced. If you express contempt or vanity in your life, that energy will return to you and affect your ability to find love. Superiority will cause an opposing excess potential to bring you into realities where the universe shows you why you're not superior.

Put simply, idealization and overestimation—of ourselves or others—will always lead to a level of disillusionment. They will detect it in you and it will push them away, or you will become disillusioned and unsettled.

EMOTIONAL BLOCKS

Our emotions are natural responses to the world around us, but we cannot let them take away from our ability to live a loving and joyous life. Grief, for example, is a common human emotion that can be incredibly powerful. If we allow it to, it can pull us into a reality where the grief is magnified and continues to get worse. The answer is not to avoid it—if you only lived on this earth for a short amount of time, you'd experience grief. We simply have to learn how to experience it without letting it consume us.

Fear is a similar block that we all have to overcome. I've met women who have had such terrible experiences with men that they could no longer be in a relationship with them. There are men who have a self-identity of fear that will keep them from approaching anyone. Even the alpha males are chicken and simply trying to put on a front. Understand how your fears interact with the person that you love or your beliefs about love itself.

Common limiting beliefs that I hear include, "I'll never find love," "That's just how women/men are," "If I let down my defenses, I'll get hurt," or "The other shoe will eventually drop." We ghost people who seem too good and indulge a deep distrust of true joy. Because our beliefs become who we are, staying connected to these hurts from past relationships pull us away from finding our soulmates in new relationships.

Be open about these fears and concerns. Be honest about your emotions as the reasons behind your actions.

David Buss and Cindy Meston surveyed 203 men and 241 women, asking them to list the reasons that they had sex. The number one reason was attraction, but then they moved (in order) through the experience of physical pleasure, to show affection, to please a partner, to harm another person (such as inspiring jealousy), to enhance social status, to express love, to return a favor, to feel a connection to their own body, and out of duty or under pressure.[77]

Opening yourself and your partner to the emotions behind your reasons is broadcasting those emotions. When we send it out into the world, we're more likely to find someone open to that same thing. Obviously, harm is a motivation that stems from anger. A sense of duty might come from guilt. To find real love, we'll need to step into higher emotions that will attract those frequencies.

As Bob Marley said, "You open your heart to knowing that there's a chance it may be broken one day, and in opening your heart, you experience a love and joy that you never dreamed possible."

Don't be afraid of love. Don't be afraid to walk up to someone. Start conversations in Target. Go get someone's phone number. Even if you're in a relationship with someone, do it as practice. You can even tell them that's why you did it. Looking at other people and opening yourself up to the possibility of joy changes your interactions. Sometimes serendipitous relationships are the most powerful, emotional, and wonderful. The only way we find them is to be open, in spite of our other emotions.

WHAT MANIFESTS LOVE

The million-dollar question is whether we can actually

---

[77] Cindy Meston and David Buss. "Why Humans Have Sex." *Archives of Sexual Behavior*, 36 (2007): 477-507.

manifest a specific person's love. I believe we can with an incredible expenditure of energy, but also that we might not want to. Your desire for a specific person is coming from the ego. You are not allowing the universe to find the best possible partner for you. I have personally manifested meetings with specific people, but that doesn't mean they will love you or that a legitimate relationship will result from it. Expecting anything more than that would be wrong—and often isn't as powerful as we imagine it will be.

Everyone has their own free will and creates their own reality. Even if you're powerful enough to maneuver situations so that you meet and interact, it cannot make them your soulmate, a particularly good love, or worth the effort at all.

For so many other sides of manifestation, being specific is important. But when you manifest love, it retains its power and effectiveness when you don't imagine exactly what they might look like or who they might be. The feeling is what makes it real. It's not the person that we need, but something much deeper.

If we want to find love, we cannot control each other in any way.

I have found six human needs that are met with love: certainty that someone will always love us; variety from day to day; significance and to feel that the person who loves you sees you as important; connection to each other; growth and becoming more; and finally, contribution. You might visualize the way those needs can be met, or think about the way your current relationship needs to meet the needs better. Understanding the nature of love keeps us on the side of love instead of obsession and defines the future.

The way the subconscious treats love is the way we will receive it. Believe that your twin flame, your soulmate is out there. Be open to finding them at any moment. Destroy the ego that lives in fear, because it is incapable of love. Meditate on this nature of love and understand the patterns that you've slipped into.

Imagine what it is to be with a person you love deeply—to eat dinner with them, cuddle on the couch to watch a movie, drive in the car with them. Set a place for them at the table, even if they aren't there. Fix up your house like you're ready for them to be there. Imagine the situations you might find yourselves in, find the feeling of the beauty of love itself. Prepare for your soulmate, because you know they are coming, and let the universe mold that interaction for you.

LOVE AT FIRST SIGHT

It can be tempting to write off the idea of love at first sight, but let's look closer at what it implies. The seemingly intangible sexual chemistry that triggers between two people—pheromones, unexplained connections, and that sense of liking someone without knowing why—has a biological source.

Remember that the brain is taking in 400 billion bits of information at any given moment, but only 2,000 of those bits are consciously processed. No matter how many mindfulness meditations or therapy sessions you might attend, you will never be conscious of all of the information your brain takes in. But not being aware of it doesn't keep any of the information from having an impact. That information is still being processed at astonishing rates, and your romantic and sexual preferences are no exception.

These subconscious measurements are part of an intuitive skill developed over millions of years—the construction of a template of everything you desire most out of a relationship. These "love maps" vary in nature. One person's map might direct them toward short-haired brunettes, while another's might include buxom redheads. Love maps lock in physical and emotional types, as well as atypical sexual needs and behaviors. These are not just conscious preferences, either—these maps are said to be mostly locked in by the age of seven. Your brain has written up a casting call for your ideal partner, and there is usually only one brief audition. The brain typically sizes up potential mates within three minutes of meeting.

The subconscious is processing information on levels that we can't consciously understand. In 2008, researchers discovered an almost imperceptibly tiny olfactory receptor called Nerve Zero, which they believe processes pheromones. These fibers start in the nose, completely bypass the olfactory cortex, and move right into the sexual centers of the brain. This makes your partner's scent a huge factor in your attraction, even when you aren't conscious of a smell.

Each person's smell has their own genetic *major histocompatibility complex* (MHC), which plays a fundamental role in your immune system. Because family members share similar genes, they often share similar immune systems. We may be unconsciously seeking out a mate with a different MHC to avoid falling in love with family members, while pregnant women are drawn to people with a similar chemical makeup during the time when they prioritize the safety of a familiar tribe over sexual needs.

Of course, all of the poets and artists in human history have not been praising chemical responses. There is a sacred, intangible beauty about the way two people are drawn to each other and fall in love. The shiver of energy, the immediate sense of recognition, the magnetic pole, the electric excitement, the novelty and discovery—love is spiritual. But if you're struggling to believe that connection can happen "at first sight," these biological explanations can help you to open your heart to the possibility.

SEARCHING FOR SOULMATES

I believe there are people who best interact with you, fulfill your needs, and help you grow. We all have a different idea of what exactly a soulmate looks like, though we can typically settle on the idea that it's the person who is meant for you.

We each have a unique set of preferences and standards. What is completely acceptable to one person in a relationship can become a deal breaker for another, including within that "soulmate" relationship. If you think that a soulmate means having everything you expect in a relationship without any compromise ever, you'll never find it. Compromise is part of growth as a couple and as individuals. At the same time, if being with a particular person means the compromise of one of your core values, that's also not likely to be the relationship that you need.

One way to clarify these values is to make a soulmate list. The clearer you are about what you're looking for before you meet people, the easier it will be to recognize a soulmate when they come into your life.

This is where we can become specific in our search. Allow yourself to be detailed in your search at first, then decide which characteristics are deal-breakers and which are not. Think about the aspects of your life that you look forward to sharing with a partner, the things you look forward to doing together, and how you feel in their presence. Ask yourself how you want to feel when you wake up next to them, what kind of lifestyle you want to lead, whether you have children, and so on. It's like typing a keyword into Google—the more specific you are, the more likely your search will return results.

If I tried hard enough and expended enough energy, I could probably decide to meet Angelina Jolie at some point in time. If you're powerful enough and creative enough, you might even be able to manifest a loving interaction with a specific person. But that's a narrow way to approach this. You might not be meant for each other in the way that you had idealized. You might find that you had only manipulated that person out of the reality that they were moving toward. You might become like a pendulum in their lives.

The more specific you are about their personality, the more powerfully this works. Create a list, filling pages and pages of qualities that you hope to find in a partner. I believe that a giddy excitement bubbles up in the universe when we create these lists.

Have a specific type of person in mind and a specific life that you want to build with them—but let the universe fill in the gaps.

EXAMPLES FOR THE SOULMATE LIST

Abundant, adorable, affectionate, articulate, beautiful, bubbly, charismatic, creative, considerate, emotional, not emotional, endearing, enjoys certain hobbies, family-oriented, flexible, fun, generous, has available money/time/affection, has great relationships with family or children or ex-spouses, happy, healthy, honest, independent, intelligent, loving, nurturing,

> playful, sensuous, sexual, smart, spiritual, successful, supportive. They cook, play golf, bungee jump, skydive, collect candles, like baseball.

IT COMES DOWN TO LOVE AND TRUST

Our understanding of the universe rests in quantum physics, and within that paradigm we know that particles become entangled. What greater entanglement is there than falling in love with someone? That bond is formed through trust, and trust is inspired when someone has your best interests at heart.

From the beginning of this book to the end of it, the thread has been and will continue to be choice. This is not a spiritual awakening but a holistic transformation.

As we become more powerful in our ability to create reality, we can begin to choose love over fear. Finding love in this moment is a choice. Finding fear is a choice. Even if you never find your soulmate, you are responsible to choose love and joy and beauty in this world. When we do, we can create a reality beyond anything we can imagine.

PART IV

HACKING YOUR REALITY

CHAPTER ELEVEN

HOW DO WE CONSCIOUSLY SURF THROUGH PARALLEL REALITIES?

For a significant part of my life, I have believed my purpose was to clarify what happened to me on the night of the home invasion. There were too many significant differences in my reality to ignore, and I had started to question my sanity. I frequently wondered if I had simply gone insane.

Could I find a technique to continue exploring parallel realities? *Even if it takes a souped-up Delorean, tell me what I need to get!*

When bizarre events happened around me, the scientist inside of me spoke up. There had to be a scientific reason for what I was seeing, even if that meant post-traumatic stress disorder or delusions of some kind. Every part of me wanted desperately to believe that I wasn't insane—and there is a part that still does.

Outside of clinical explanations, I read everything that I could. What I had gone through was transformative, and there were people out there with explanations for that transformation. While I had already done a fair amount of research before the home invasion, there's only so much you can accomplish before an experiential basis is required. You need to live it to understand it. Once it seemed I had experienced a reality shift, I went back to read even more. With some understanding of a theory around parallel realities, I could see just how deep and complicated the scientific reasoning behind it all could go.

In part one, we talked about some of those theories and explanations that define my experience and techniques. In part two, we looked at the practical ways that we can take control of our present reality. Now, it's time to start hacking reality. Not to just move into another reality, but to do so in a way that can significantly change your life for the better.

Let's begin with an arsenal of tools that you can always have at your disposal.

COMMON SENSE TECHNIQUES

Move. There's nothing mystical about this, but it's powerful. Think back at the times you've moved to another house, another town, or even went on a long-term vacation. You're literally moving to another reality. The people you hung out with in that new place, the food you ate, the air you breathed—everything completely changed.

I'll be honest and say that one of my fears is moving. Some people are fearful of fires and snakes. I'm afraid of moving. Every time I've done it, however, it has been an incredibly transformative experience. The fear is often wrapped up in how tightly we hold onto our possessions, which is also the liberation of it. When you move, you start to let go of the things that you've attached yourself to.

If you want to change your reality right now, you can do it. Just choose to move.

Learn a new language. The syntax and linguistics that we use define our reality in many ways. Some languages break everything into gendered terms or use words in unique ways, and that defines the way their speakers think about and understand the world. When you learn a new language, you gain an entirely different perspective than the one you're accustomed to.

Change your name. Brian Scott is a pseudonym for a last name that people can't spell—Bengtson, which I have to explain how to spell every single time! When I chose to change my last name, it carried more power than I realized. Your name has a lot of your reality attached to it. If you can change your name, you can actually become someone else.

PHYSICS SOLUTIONS

While these may not be practically useful, there are several ways that physicists have theorized could be used to travel into parallel realities.

Wormholes and black holes. Theoretically, wormholes could eventually be a way to travel into other realities. These are portals, which connect to other universes. Many physicists theorize that black holes lead to other universes as well. Our own universe might be the other side of some other universe's black hole. Someday, perhaps, we'll find the way to travel safely through these portals and they will become hacks of their own.

For now, that kind of travel isn't an option—and it might not be necessary. When we understand the power of consciousness, parallel realities are much closer than we think.

Quantum entanglement. Physics tells us that entanglement is a form of connection. If we become entangled with particles located in realities outside of our own, it could explain intuition and understanding that we otherwise wouldn't have. Entanglement may pull us into other realities.

Folding space. Some physicists believe that we will, at some point, be able to fold space to minimize directions and move entirely into other areas of reality.

Warp Drive. You may laugh, but have you heard of the Alcubierre Drive? Originally posited by physicist Miguel Alcubierre in 1994, this concept gets around Einstein's equations for limitations due to the speed of light. Some physicists are now saying that this might actually be feasible.[78]

Theoretically, the Alcubierre Drive achieves faster than light travel by stretching the fabric of space-time in a wave (sound familiar), causing the space ahead of it to contract and the space behind to expand, a spacecraft inside this wave would be able to ride in a warp bubble and move beyond the speed of light. My theory is that we created this possibility by tapping in to source and finding a solution. In an interview with Frederick Dodson, he explains the exciting idea that all science fiction is real in some part of the universe, that these ideas exist, all imagination is real.

SUBSTANCE HACKS

Psychedelics. Without advocating the use of psychedelics, we still need to talk about them. The number one thing to know about the use of these substances is that any access to parallel realities will be temporary, and it cannot be used as a crutch. However, this has been a practice for over half a century, and psychedelic science is coming back to life under the efforts of multidisciplinary associations working on legal studies here and abroad.

---

[78] J. Agnew, "An Examination of Warp Theory and Technology to Determine the State of the Art and Feasibility." *American Institute of Aeronautics and Astronautics.* Published Online, August 16, 2019. https://arc.aiaa.org/doi/10.2514/6.2019-4288.

There are experimental protocols for microdosing of nonaddictive visionary substances such as psilocybin, mescaline, LSD, ayahuasca, marijuana, and MDMA. According to James Fadimen from *Psychedelic Explorers Guide*, the microdosing protocol that he suggests has improved concentration, shut out all distracting influences, and moves people into flow more easily. In one survey of over two hundred individuals, participants ranked categories like focus, problem-solving, and creativity on a scale of one to six, with an average response of 4.5–4.76 in each.

Joe Rogan's podcasts indicate DNT or ayahuasca is a cultural element right now. People are traveling to Costa Rica to use this drug and experience incredible transformations. There are theories that DNT may activate the pineal gland. For a transformational story about peyote, you can read Jerry Spence's book *Making of a Country Lawyer*. And of course, Carlos Castaneda has written extensively about his experiences with psychedelics as well.

We could dedicate a whole book to the use of drugs to access source and push the recipient into alternate realities. It's a practice extending back to the beginning of time, when humans walked off into the plains and ate mushrooms and created worlds in their mind. With what we understand about choices and realities, those very experiences could have transformed us entirely—it's possible that we are who we are because of those substances.

Nootropics. These are substances intended to improve mental performance with minimal side effects to the body. You might have heard them referred to as cognitive enhancers, memory enhancers, or even smart drugs. If not, you're probably familiar with the use of coffee or tea to become more focused in work. That's the idea behind nootropics.

Nootropics supplements are natural compounds that elevate cognitive performance to peak levels, keeping you alert and focused for a long time, without the jitters. Studies indicate boosted memory, increased confidence levels, more personal motivation, and better overall brain function. Each of these things can pull you into better parallel realities by improving the thoughts that you have.

Common nootropics include ginkgo biloba, St. John's wort, glutamine, acetylcarnitine, dimethylamino bitartrate or DMAE, vinpocetine, and more recently, keto ethers, which Russel Brunson uses when he speaks at large conferences. These substances and compounds are being used more and more as we become aware of biohacking potential, and they are certainly worth looking into.[79]

SENSORY MANIPULATION TOOLS

The Ganzfeld effect. Humans have been aware of this effect for thousands of years. Some meditation techniques

---
[79] More biohacking techniques can be found in the back of the book.

include staring at a blue, cloudless sky until there's no sky, or at a candle until nothing in that room exists except that candle. The yoga meditation *Tratakum*, or steady-gazing, involves concentration on an unchanging or external object such as a rock or a mandala. Michael Hutchinson says that the blank out effect seems to be a key to many of the well-documented benefits of meditation, ranging from stress reduction to increased sensory acuity and mental clarity.[80]

In his research, Robert Ornstein discovered that a blank-out like this was not merely the experience of seeing nothing, but a complete disappearance of the sense of vision. Those who experienced a blank-out, for instance, didn't know whether or not their eyes were open. Institutions have concluded that periods of monotonous stimulation indicated a similarity between continuous stimulation and no stimulation at all, such as in the repetition of a single word or phrase as part of meditation as nearly all spiritual traditions employ.

For a long time, researchers have tried to induce this state in the laboratory. Techniques have ranged from dense, uniform fog to translucent goggles. On Amazon, you can order the Covena Deep Fishing Ganzfeld goggles for about a hundred bucks, or you can do it yourself. One simple, DIY way that you can induce a Ganzfeld effect—or perceptual deprivation/a blank-out—is to cut a ping-pong ball in half, put a slit in each, then put them

---

80 Hutchison, Michael. *Mega Brain* Power (CreateSpace, 2013).

over your eyes. Shine a light directly onto the ping-pong balls and keep your eyes open. With a relaxed, soft gaze at a uniform, evenly illuminated but featureless visual space, it's much easier to settle into Ganzfeld.

Float tanks. Sensory deprivation is a practice that has led to great miracles of self-regulation for thousands of years. From Hindu firewalkers to Tibetan monks sitting in subzero snow wrapped in icy blankets, to yogis who bury themselves alive for days in airtight boxes and healers who puncture themselves with knitting needles, the loss of sensory input seems to be connected with a greater control of reality.

Where goggles can create a total blankness in visual space, float tanks can deprive senses in complete blackness and loss of sensory input. It effectively produces a blank-out directly by shutting out or dramatically reducing most external stimuli, including sound, gravity, and tactile sensations.

Float tanks are powerful ways to come in touch with your mind and understand the very essence of reality. Ganzfeld and float tank experiences direct attention inward by giving the reticular activating system nothing more than an unvarying stimulus. With no outside feedback for the RAS to process, you gain access to the open wavefront of energy that defines reality itself.

If being in the tank increases your awareness of the most minute internal processes, then the increase in awareness once you leave the tank is no less profound. People who emerge from a float tank are often delighted to find that the world seems to have changed. They describe the world as fresh, glowing, illuminated, bright, intensified, vivid, and luminous. When you cut down input to your senses by going into the tank, your senses seem to respond by expanding and becoming more sensitive.

If we see the mind and body as a single system, then it becomes clear that external stimuli are constantly working against the system's equilibrium. Every noise, every degree of temperature above or below optimal levels, every encounter with other people, every feeling of responsibility, guilt, desire—everything we see and feel incessantly interrupts our self-contained system. All current evidence indicates that floating influences chemical secretion in the brain and affects every aspect of the user's behavior, including moods, emotions, and response.[81]

> The first few times, you'll have to be careful to not touch your eyes after having your hands in the saltwater of the tank. If they have goggles, I recommend using them. At first, a lot of people don't know where to put their hands, so they'll put them above their head until it becomes uncomfortable, then put them down lower. Once you get relaxed, gravitational muscular tension will allow you to detect smaller sensations than

---

[81] Philip Applewhite, *Molecular Gods: How Molecules Determine Our Behavior* (Prentice Hall Direct, 1981).

> usual—in other words, it intensifies the sensations that you otherwise overlook.

Bathtubs. I don't believe it's a coincidence that famous people all have wonderful bathtubs. Different than floating, a bathtub creates deep relaxation in its own unique ways. They say that spirits avoid the water, and it is historically understood that water is cleansing and powerful. The vibrational blank slate of the water itself can amplify your intentions. Salt and bath bombs can help create a relaxed state—your very own tub of holy water—which becomes an amplification effect. In the relaxation of a bath, you can move into realities, discover things, and manifest things that are likely far beyond what you expected.

Water and temperature training. The CEO of Twitter, Jack Dorsey, says that he jumpstarts his morning with an ice bath and ends his evenings with a sauna and ice bath. He explained that his evening hydrotherapy routine looks like fifteen minutes in the barreled sauna at 220 degrees followed by a 37-degree ice bath for three minutes. He completes this cycle of extreme three times, then finishes with one more minute in the ice bath. He said that going into an ice-cold tub straight from a warm bed "unlocks thinking in my mind, and I feel like I can will myself to do the thing that seems so small but hurts so much. I can literally do anything." Tony Robbins does this too—every morning he jumps into a cold bath.

In my own experience, temperature training is powerful, and I have used the Wim Hof method. Wim Hof set the world record for the longest time in direct, full-body contact with ice. He has done this sixteen times, and has developed a technique for training. There are several variations, usually involving controlled hyperventilation with thirty cycles of breath. As you hold your breath longer and longer in these cycles, your body warms up and is better able to go into the cold water.

Being in a super cold lake or bath really is transformational. You gain better control of your mind and body. The stress of extreme exposure to cold is good for health, athletic endurance, preventing muscle atrophy, increasing neurogenesis, improving learning and memory, and improving longevity. Cold showers can be used to treat depression, releasing norepinephrine into the brain to affect diligence, focus, attention, and mood. It also has a role in pain, metabolism, and inflammation, as temperature response is mediated in the fight or flight sympathetic nervous system. More specifically: norepinephrine can rise 200–300 percent with cold immersion near zero degrees Celsius for twenty seconds.

The cold activates thermogenesis, which is the way the body produces heat. It activates brown adipose tissue, which is the form of fat that burns regular fat through oxidation. It has a mitochondrial effect, and the more mitochondria there are, the better aerobic capacity we have. Similar effects have been observed with heat shock proteins, and the combination of both heat and cold is incredibly powerful.

If you do temperature training, avoid alcohol and drink lots of water. Don't do it if you're sick. Don't stay in saunas for a long period of time. Cool down gradually and don't overdo it. I had kidney stones after doing temperature training, and it might have been because I wasn't drinking enough water.

Light stimulation. The color of light measurably, significantly affects neurochemical access that changes our mindstate and thoughts. In one study, when subjects were measured for neurochemicals and hormones before and ten minutes after a single, twenty-minute exposure to violet, green, or red lights at a frequency of 7.8 Hz, Shealy found that the exposure produced significant increases of 25 percent or more of key hormones. This included growth hormone, luteinizing hormone, and oxytocin, which all help promote affection, love, sexual hormones, growth, and increased fat metabolism.[82]

Among recent developments for light-color therapy, there is the Lumitron, which was invented by John Downing. It uses Xenon gas as a full-spectrum light source, sending light through one of eleven different color filters that can be changed out. The spectrum ranges from violet, indigo, and blue on one end of the spectrum to red and orange on the other. Then the colored light is

---

82 C.N. Shealy, et al, "Effect of Color Photostimulation upon Neurochemicals and Neurohormones." *Journal of Neurological and Orthopedic Medicine and Surgery*, 17, (1996)95-97.

delivered to the user's eyes through a small circular lens. Downing says that the proper color stimulus will reset the biological clock and create more balanced hypothalamic discharge rates.

More broadly, this understanding of color and light changes implications around the use of color in a given environment. Looking at art, for example, or specifically using certain colors can change the way your brain works and alter your reality.

Acoustic field generator. An increasingly popular "alternate mind machine" combines a variety of stimuli to provide a whole-body, multisensory experience. It generally includes light, sound, and physical vibration, and as the name suggests, the key stimulus is sound—not just heard through the ears, but felt as vibrations throughout the body.

One way these devices work is by stimulating the release of pleasurable neurochemicals. Dr. Abram Goldstein, head of the addiction research center in Palo Alto and professor of pharmacology at Stanford, found a link between musical thrills and increased endorphin production: shudders of ecstasy are produced by emotionally moving music.[83] When you're lying on a sound table, the powerful information and emotion that's transmitted through musical vibrations is processed in the brain and creates that neurochemical experience.

---

83 Avram Goldstein, "Thrills in Response to Music and Other Stimuli," *Physiological Psychology*, Vol. 8, 1, (1980). 126-129.

Acoustic field generators can be as complex as a sophisticated sound processing system inside a state-of-the-art computerized dome that incorporates modulated light-color goggles that flash in sync with the music. These systems can cost as much as $100,000. However, a massage chair in combination with moving music and light-color inputs can be just as powerful.

Virtual reality. Meditations using virtual reality may be the new cutting-edge sensory tool. The idea would be to have a 360-degree video of beautiful places combined with a guided meditation. I believe it will produce that Ganzfeld effect and make meditative practices that much more effective. At the Advanced Success Institute, we have developed several revolutionary meditational systems using virtual reality.[84]

Motion systems. A number of motion systems on the market right now include reclining chairs that revolve, and beds that tilt, revolve, and rock. These systems are being used clinically for such purposes as treating brain damage, learning disabilities, and drug addiction. Motion systems are popular in brain-mind gyms around the world, and there's evidence that they alter brainwave activity, increase alpha and theta activity, and enhance hemisphere synchronization.

Several new motion system models are available at relatively low prices, making it possible to purchase them for home use. I have a Potentializer, which creates a rock-

---

[84] www.advancedsuccessinstitute.com

ing motion that feels like you're being rocked like a baby. It moves the fluid around your vestibular system and inner ear, sending signals to the brain that flood it with electrical stimulation—like exercising the brain directly.

If you don't have an actual system, EEG results and other evidence shows that spinning has a profound optimizing effect on the neuro-efficiency quotient, which is a measure of how rapidly electric signals are transmitted from one part of the brain to the other. If you find yourself in a situation that requires peak mental performance, take a break and spin around at your desk for a few minutes every half hour or so. One early study cited by the *LA Times* says that this kind of motion has "increased people's IQs, changed brainwaves in adults and learning-disabled kids, turned C students into A students, and affected cognitive function in autistic children."[85] This is also the first part of the five Tibetans exercise mentioned earlier.

TECHNOLOGY TOOLS FOR THE REALITY SURFER

Brain simulator headsets. These devices manipulate the way the brain works in very subtle ways. One that I've used is called the PlatoWork, which is a plug-and-play device designed for neurostimulation. It looks like a pair of headphones, but one set lines in the front and back of your head.

---

85 Connie Zweig, "Mind Gym: Exercise for the Brain," *Los Angeles Times,* December 1987. Accessed online, https://www.latimes.com/archives/la-xpm-1987-12-01-vw-25909-story.html.

There is a science behind these devices. Each of the billions of neurons in the brain communicates with the other, and they send signals when they are activated by another incoming signal. That creates electrical changes that pulse through the brain. As you read this book, nerve impulses are running to the neurons and facilitating communication all across your brain. Brain stimulator headsets, or transcranial direct current stimulation (TCDCS) devices, make it easier for the neurons to be activated. More nerve impulses are transmitted, and thereby, according to research, brain activity is increased.

This stimulation allows you to focus your attention. It boosts concentration by minimizing impulses from the reward-seeking regions of the brain and mind-wandering in a noninvasive way. It is used to modulate neuronal activity and increase synaptic plasticity through this low intensity current from the headset to the scalp.

Vibration gravity. As we walk around throughout our day, our energy is constantly, gravitationally pulled downward by the earth itself. When we discussed the Bindu chakra on the back of the head, we talked about doing headstands or using gravity boots to reverse the flow of energy and activate that particular chakra.

One powerful way to enhance that practice is to stand on a vibration machine for a minute or two first. If you follow that with a gravity machine or gravity boots, the energy that you just vibrated into your body comes down through your chakras to activate that Bindu connection to the source. In my experience, this can lead to very powerful manifestations.

Note that this can be dangerous if you have high blood pressure, and it can cause pressure on the ovular area of your eye. Talk to your doctor before trying it.

Neurofeedback. Monitoring the way your brainwaves work can expand your ability to move through consciousness, and this technology is becoming increasingly more available. If you don't have the money to go to a 40 Years of Zen event, you can purchase headsets like the Muse or the NeuroWave Mobile 2. These devices give you access to the brainwave states that you're in by gathering feedback from your brain and the way it's working. When you know what state you're in, you can gain some more control over it to move into parallel realities and expand your consciousness.

ADVANCED REALITY HACKS

Magic hacking. There is a long history of books about magic that stretch all the way back before Christ was born, coming from all different cultures. A lot of these magical activities and rituals and behaviors can be hacked, because they are often exploring the scientific nature of maneuvering through parallel realities. They're creating vibrations and tapping into what we now know are powerful techniques.

The more we learn about magic rituals, the more we

see that they work. One example is the practice of creating representational symbols of magic spells. It turns out, this is a powerful method for reality hacking. Dr. Joe Dispenza does something similar—by creating a symbol for the thing that you want, you endow it with your power. Keeping it around keeps that power around as well.

You can create these symbols as elaborate pieces of art or as simple pictures. I've created one that enhances my ability to relax, and I look at it for thirty seconds before I meditate. It changes my mindset and enhances the meditation. There's an interactive play that occurs between my intentions and the symbol.

Another popular concept in magic is the idea of chaos magic—that no particular thing is magic, but that it can be created all around us. Perhaps that's true. Even if there are mystical or crazy-sounding practices to wade through, there is some truth in everything, and it's worth exploring.

Guided quantum jumping. In some of these deeply relaxed states, you truly do go into the void. Unfortunately, a lot of people will come out of them saying they were so relaxed they didn't know what to do. Dolores Cannon created the QHHT—Quantum Healing Hypnosis Technique—to guide intentional actions in that state, where she is able to bring up people's visions of past lives going all the way back to old times.

Using a specifically designed guided meditation following an energy exercise may be the most effective technique.

If you have a particular intention around something transformational in your life, time in the void can lead to quantum jumps. Find someone you trust who can guide you, or follow a guided quantum transformation meditation. I have more than forty different meditations that you can play with on my channel.

Astral projection. This is not necessarily the same as traveling through parallel realities, but a way to attune yourself to your energy body. It's projecting yourself outside of your body, or an out-of-body experience. Some people feel as though they have looked down at their body while lying in bed. Others have been in different parts of their house, they've seen or heard things happen that they shouldn't have otherwise known. Many people don't know if the experience was "real" or not, but studies favor it.

Astral projection is not only a real technique, but it is something that you can train yourself to do. It teaches you to become attuned to your own energy body and to switch bodies as well. If you're interested, Bruce Mercer's *Mastering Astral Travel Projection* is extraordinary.

Frequency shifting. Frequency and vibration are different, though frequency is tossed around a lot when talking about spiritual and metaphysical things. Sometimes, it's just used as a code word. Really, frequency is the *rate* of vibration. It's the oscillation measured over a period of time, usually a second. It's a repeating sequence. A heartbeat, for example, has an average frequency of sixty to one hundred beats per minute. Frequency is the cycle of waves measured in a sequence, and vibrations are the contraction of energy within that frequency, while oscillation is the expansion of the energy.

We literally define our reality with each action, thought, and word. We have already talked about how, when we shift to a higher frequency in our personal energy signature—vibrating on a higher level—we attract more positive emotions and experiences that match the higher frequency. Imagine attending a large party and gravitating toward people you feel a connection with. This is how frequency works. Love, for example, is a high-frequency emotion. Ego-based mindsets are lower, and attract negativity, stress, anxiety, and depression. This awareness changes the way we think about life, from how we decorate our homes to what we eat and who we associate with. Understanding frequency changes the way we attempt to maneuver through realities.

Fractal animated meditations. If you were to look at the patterns of energy that contain all possibility—the psi-tronic wavefronts—I believe they would look like fractal patterns that are constantly moving and alive. Dr. Joe Dispenza, again, talked about them as kaleidoscopes:

> When I'm having a mystical experience, it seems more real to me than anything I've ever known in my life, and I lose track of space and time. Often, just before I become entwined in it, I see in my mind and sometimes in my outer

world, circular geometric patterns made of light and energy. They tend to look like mandalas, except they're not static. They are waves of interfering frequencies that appear as fractal patterns. The only way I can describe these patterns is they are alive, moving, changing, and ever-evolving into more complex patterns.

I have seen these patterns as well. There's something in this imagery that goes beyond relaxation and into an ideal brainwave state. While Dr. Dispenza uses kaleidoscopes, there are a number of different places on YouTube where you can access meditations built around moving fractal animations.

Even more effective than kaleidoscopes is the fractal zoom, the infinite nature of this has led to multiple reports of powerful transformations.[86]

It has become an art and a science. By looking at this kind of imagery, your mind reaches a state of access to the wavefront. It puts your brainwave state into alpha and theta, where you're open to possibilities. If you combine this practice with a mind movie, I believe you'll see incredible transformations in your life.

The As If method. If we have an understanding of parallel realities and we have access to future realities based on what we focus on, then we know that we can move toward the futures we want through how we think about things.

---

[86] For a three hour Mandelbrot zoom, you can go here: http://www.therealityrevolution.com/understanding-how-to-unify-the-heart-and-mind-in-reality-transurfing-ep-109/.

The idea is that acting "as if" something has already happened, we can create those realities.

According to William James back in 1884, the notion of behavior causing emotion suggests that people should be able to create any feeling they desire by simply acting as if they're experiencing that emotion. He said, "If you want a quality, act as if you already have it."

A few years ago, scientists tested this concept by putting participants in a brain scanner and asking them to contort their faces into a fearful expression. Unlike psychological studies of the past, these participants did not have to tell the researchers how they were feeling. Instead, the brain scans told the story. The faces created highly active indications that the participants were experiencing genuine fear—even though they were simply contorting their expressions.

You might have noticed this phenomenon if you've ever smiled and then started feeling happier. Just as you can create a happy response in your body by making a happy face, your behavior influences how you feel. If you sit up straight and hold your body confidently, you start to become more confident. When actors on a set behave as though they love each other, they often end up in a relationship after filming has ended. If you're struggling with speaking, act as if you are a speaker and your brain will change.

Maybe you're accessing some other reality where you are happy, in love, and a speaker. Whatever is happening, "acting as if" is a simple thing to do that we rarely think about trying.[87]

Identity shifting. One of the things that I have found helpful in coaching is to understand the concept of identity. Our identity defines everything about who we are, and the fun practice of creating a new identity can create a new reality as well. There are plenty of hard science indicators that an identity shift causes a reality shift, but one of the more extreme examples comes from Frederick Dodson's *Parallel Universes of Self*.

In researching multiple personality disorders, it has been proven again and again that extreme changes can occur when someone shifts identity. People have lost twenty pounds within a few days, facial scars disappeared completely, and voices, memory, and looks transformed within minutes. Identity shifts do not change the body alone, but I believe science will one day admit that it changes the entire body of experience.

The outside world is not separate from consciousness. This is similar to "as if," but more extensive. This is about your entire identity. If you say you are a writer, in more than just a moment, you're going to think about being a writer, you're going to actually write, and eventually you're going to have to let go of your past identity that was not a writer.

To create an actual reality shift, go back to the old identity. Go to places you would have gone as that person, go to

---

[87] I recommend *The As If Principle* by Richard Wiseman.

them as the new person, and start to differentiate between them. Give yourself the power to choose your identity—including your name, profession, activity, family, friends, foes, home, possessions, documents that prove identity, things that prove your past, pictures and certificates, languages and dialects you speak, words you use, intentions, idioms, the way you look, and the gestures you use.

A lot of times, the spaces we enter define our identity. Instead of allowing them that power, think of them as a movie set that you can choose to participate in or not. Become fluid to your surroundings. Choose sets that align with the identity you want. If you want your identity to be a boxer, go to the gym and start boxing. Watch videos. Notice your body feeling different and becoming who you want to be. Allow your new identity to define you, to program your thoughts, and to shape your reality.

Luck training. Richard Wiseman wrote a book all about luck, and he identified four categories of luck training that can move us into new realities.[88] The first one is to maximize your chance opportunities. People who are lucky have specific habits. If there are infinite realities available to us, these particular behaviors seem to open up those possibilities. The opportunities include networks, conversation initiation, connections, openness to new opportunity, and a relaxed attitude.

The second category of luck training is to listen to luck hunches. Lucky people listen to their gut feelings

---

[88] Richard Wiseman, *The Luck Factor*. (Miramax, 2003).

and are in tune with their intuition. The third is to expect good fortune. Yes, lucky people have bad things happen to them, but unpredictable events work out in their mind. They expect good luck to happen, and they have behaviors and activities that align with those expectations.

Finally, lucky people turn bad luck into good. They see the positive side of bad luck. This is counterfactual thinking—against what the world is telling you, you don't have to be right all the time. Just realizing that and embracing it can bring good luck into your life. Use a luck journal, write down times when you've felt lucky, identify these categories of training in your life, and gather information about the way luck works around you.

Write out your reality. If you want a particular reality, write it out exactly. A lot of writers by trade will tell me that the things they write as a story show up in their lives as crazy coincidences. There is something powerful in the act of writing, just as there is in the act of reading. Every single book can be an exploration of another reality. You can see how someone lived an entire life, fiction or nonfiction, in the span of hours. Read, listen to audiobooks, and expose yourself to other lives, habits, and behaviors.

No one else needs to read what you write in this exercise, however, so don't worry about making it a book. Just be specific. How do you wake up in the morning, what do you eat for breakfast, and what kind of car do you drive? Write out all of the things that you want, and incredible things will start to change.

PERSONAL HACKS

Multimedia vision boards. The traditional vision board is powerful because it's a constant reminder of our goals, even when we're thinking about other things. We see images of the things we would like to manifest, and according to *The Secret* and countless individuals, these boards have led to incredible experiences. With new media discoveries, we're starting to advance to new levels. Seeing movement and action connected with your vision has a powerful effect that a board alone can't match.

One company is working on creating virtual reality vision boards that depict things like riding on the yacht that you otherwise would have written about or found a static picture for.[89] Until then, in *Becoming Supernatural*, Dr. Joe Dispenza talks about creating mind movies. *The Mind Movie* app allows you to create playlists and combinations of videos that become a way to immerse in your vision. You can also find mind movies that other people have created on YouTube.

Chronobiology. When I first began looking into my own rhythms, I didn't realize I was stepping into the field of chronobiology—the interplay between biology and time. Scientists have long observed the way humans are influenced by a variety of biological rhythms, such as the yearly rhythm of seasons, monthly menstrual rhythms, sleeping and waking rhythms each day, and waves and cycles of rhythms throughout the day such as

---

[89] Link to product: http://www.visionboarddr.com.

hunger and arousal. While we're familiar with circadian rhythm, these latter daily patterns are called ultradian, from the Greek concept of *ultradize*, meaning "beyond daily/many times a day."

Just in the last few years, chronobiologists have made astonishing discoveries about ultradian rhythms and their influences on our bodies and minds. For example, our bodies are genetically programmed to operate on 90- to 120-minute rest/activity cycles. If you've been working hard on something for hours and feel like you're struggling, you might need to take a break to satisfy that need for rest. These rhythmic patterns to the day pull us back into the reality that we're in—which means we may get sucked back in when we're trying to reach an incredible goal. Awareness of these rhythms can help us avoid slipping into subconscious patterns.

It's not just activity that comes in waves, either—we're oscillating between left and right brain strengths throughout the day as well. Neuroscientist David Shanahan Kelsa, of the Soft Institute for Biological Sciences, did EEG studies of both right and left brain activity simultaneously, and he found that dominance shifted back and forth in a wave-like rhythm. The average time for the cycle? One-hundred-twenty minutes, just like the basic rest/activity cycle.

There are different times of the day when we're able to have spoken and spatial abilities. In peak performance studies, peak performance occurs every ninety minutes.

Tests on mental alertness, complex tasks, creativity, suggestibility, receptivity, optimism, pessimism, rest, and healing all take place in these cycles.

If you're feeling trapped in a timeline because of these rhythms, start looking at your clock. Speed up your behavior or slow it down as the cycles progress. Feed into your day more intentionally and utilize those rhythms to dominate or change what you want to do.

The two-cup method. This is another simple but powerful method that you can bring into your practice. Essentially, pour two glasses of water, grab a pen and a couple of Post-it notes. If you want to increase the amount of money you have in your checking account, for example, on one cup you will write the amount you actually have in your account. On the other cup, write down what you want to have.

Now, hold the first cup in your hand while you imagine the second cup's scenario in your mind. Picture how it would feel and what it would be like, then pour the water from the old cup into the new cup and drink it all.

In discussions about this exercise on Reddit forums, people have been shocked, even disconcerted, about what comes true. Many experiment with prosperity or weight loss or relationships, and when they succeed they tell others to relax and just try it themselves.

My personal theory about this exercise is that it works when you're able to replace the water in your body with these intentions. If you were to continually do this over and over again, as a regular habit, your body will eventually be filled with the visualizations of your future. I believe it could have changed my outer reality—after all, we have established the power of water. It carries the vibrations that we give it, and if we can fill our bodies with these intentional, high-frequency vibrations, that is what we will become.

Mindfulness training. People sometimes hear the word mindfulness and think of religion, but it's more of a biological process. Mindfulness awareness techniques train the mind to focus on the moment-to-moment experience. It is about brain hygiene, not faith, though various religions may encourage this level of health through the practice of mindfulness.

Learning this skill is a way of cultivating what we have defined as an "integration of consciousness." In a study on obsessive-compulsive disorder done at UCLA, mindfulness was used as a component of the treatment, appearing to be effective for adults and teens who struggle to pay attention. Prefrontal regions of the brain are thicker in those who practice mindfulness—an awareness of the body, sensations, and movement activates neural circuits in that part of the brain.

But long before studies on psychotherapy, mindfulness has been a practice for 2,500 years. When we have attunement, either interpersonally or internally, we become more balanced and regulated, and this has been true for the whole of human history.[90]

---

[90] Find a guided meditation for mindfulness training at https://therealityrevolution.com.

Creativity challenge. For thirty days, create something new every single day. It could be a stick figure drawn onto a piece of paper, a new dance, a new phrase to sing, a new poem, a playlist, a meal, a picture, a new exercise routine...Once you make this a goal for thirty straight days, you will discover new levels of creativity that you didn't know you had.

We're often stuck because we're trying to create a reality without being creative. Open up the part of your mind that is creative and you will gain access to a beautiful sense of the source. You will enter into flow and will likely find yourself doing more and more creative things than you knew you were capable of.

Release. Often, we're stuck in a reality because of the unpleasant thoughts, painful memories, and grudges that we hang onto. The key to manifestation and reality creation is not always in the creativity but in the release of what's holding you back. By releasing these things, we eliminate blocks that keep us from moving into other realities.

Infinite possibility. Doing something new and exciting is the key to living a fulfilling life and finding the reality that you want. When we get stuck in our ruts, our energy becomes stagnant and stale. Take matters into your own hands to open yourself up to new experiences, people, and situations—which will often wake you up to what you need.

Plan a spur of the moment trip or call an old friend for a fun rendezvous. Take a different way home or try a brand new food. Not all adventures have to be big. Just by choosing a new route or a new way of being, you gain access to countless possibilities.

In *Into the Wild*, John Krakauer said that so many people live with unhappy circumstances but will not take the initiative to change their situation. They're conditioned to a life of security, conformity, and conservatism—all of which appear to give peace of mind, but in reality nothing is more dangerous to the adventurous spirit within a man than a secured future.

The very basic core of our living spirit is passion for adventure. The joy of life comes from our encounters with new experiences. There is no greater joy than to have an endlessly changing horizon under a new and different sun. We are only stuck in our reality because we remain stuck in our reality. Embrace new adventures. Start small and see where the wind takes you.

Open your heart. If you go back and look at the people in your life, you will find people who could have chosen great job opportunities or done great things, but their hearts were closed. Opening your heart doesn't mean you'll become a doormat for other people and their personal dramas, but you will become more vulnerable and courageous in receiving love.

Unfortunately, not many of us get enough of high-frequency love. We can't afford to close ourselves off to what's available. Open your heart to people and to the incredible possibilities around you.

CHAPTER TWELVE

KEY CONCEPTS TO UNDERSTAND

REALITY SHIFTS

Have you ever had objects move inexplicably in your house? Have you lost something, only to find it in the most impossible place? You may be experiencing a reality shift. After Cynthia Sue Larson wrote *Quantum Jumps*, she wrote *Reality Shifting*. Her theory is that the shifts become noticeable when one reality dissolves into another, and a gap is left. We notice startling appearances or disappearances, much like what I experienced after the home invasion. She explains that the universe is conscious, and self-awareness is a loop of self-referential circularity. As we contemplate our own consciousness, points discontinue wherever an observer oscillates between one view and another. When we go back and forth in this way through time, in a discontinuous transition, reality shifts.

DOPPELGANGER INTEGRATION

Burt Goldman uses this term, explaining that we should give ourselves a little bit of time to adapt. When I first attempted to maneuver, it wasn't successful. When I went back to learn more, this concept came to light for me. The idea is that you're truly traveling to a parallel reality, which means your mind and body might be different. If you try to just jump in and go, your body will resist and go back to where you were. Give yourself a little bit of time in that body to adapt and integrate. Several of Dr. Joe's meditations will have you lie down wherever you're at after the void. Do something to give yourself a few minutes to adapt while still in a meditative state. In that time, move from a doppelganger of yourself into a complete fulfillment of that reality, where you believe it is real and not simply imagination.

MEMORY INTEGRATION

Dominic O'Brien wrote a book called *Quantum Memory* that helped me learn to maneuver through parallel realities. He tells the story of how he watched someone look at a full deck of cards and memorize the exact order after fifteen minutes. When he tried to figure out how someone could do that when his own memory was pretty bad, he developed the Journey Method. In it, whatever objects you memorize should be linked together with a simplistic image. Create a feeling, and connect all of your memories to it. If you need to memorize the word "book," identify the kind of book, including color and title and topic. Since we are all living out memories of our past, I integrate this technique into my parallel reality movements.

O'Brien uses a system of linking objects to each other to create a larger memory. By linking multiple ideas together, you can create more sophisticated and detailed realities with up to ten different aspects. For instance, think of how you remember your house much like walking through it from room to room connecting rooms as you go along. Connect the visualized memories together to create a complete reality with several layers at the same time.

BIOHACKING

We know that food is energy that we transfer into our bodies, and it's a topic that we could spend books and books discussing. What's interesting is that food is mostly made up of water, and water is uniquely influenced by our thoughts. We need to be aware of the intention that we transmit into our food, as a start. Even more, we can use food to directly increase energy. This is called biohacking, and its practice can increase mental clarity and help meditations.

One way is to try a thirty-day no sugar challenge. While sugar increases energy, it can also dramatically decrease it. This is an obvious first step to biohacking that can carry a lot of benefits without much effort. It can help burn fat, upgrade focus, and even increase testosterone in men.

Another way is to eat like your grandma did. On my grandmother's ranch, she would have her own chickens that she collected eggs with these deep orange-yellow yolks. She knew the guy in town who raised cows, so her meat was all grass-fed. Her fish were wild caught and her produce was pesticide-free. I personally believe that the energy used in producing food will transfer into your body, and eating better will make your energy better. This can also be enhanced by drinking filtered water and nutritive smoothies.

Another biohack outside of eating food is to use cryogenic freezing. If you can't get to a chamber, try taking a cold shower for thirty seconds to a minute. The cold reduces inflammation and creates a significant jolt of energy without ingesting outside chemicals to do so.

You might also look for your genetic strengths. For example, your blood type can have an influence on how you digest food. Tim Ferriss has a practice of blood testing on a regular basis to find out more about his body. As more people do this to learn more about themselves, it's becoming more accessible and inexpensive. There are several books you might read, as well as genetic testing that you can undergo to learn more about your body and the way it works. Learn about how your body transfers food into energy and what you need for your specific body type and circumstances.

Just as important as food, we have the ever-present issue of stress. The cortisol that stress produces reduces energy and weighs you down. Figuring out what causes your stress and reducing it will increase energy from its source, and you can also bring exercises into your routine to counter stress when it happens. While there is no particular exercise routine that you must follow to increase energy, I have found that exercise requires and involves adaptation. Within a month, your body will become used to it. Change your exercise routine every twenty-one to thirty days, and get exercise at least five days each week. Brendon Burchard recommends spending five minutes of every hour doing some kind of exercise, like Knocking on the Door of Life, from chapter 5, to get a little bit of energy that will accumulate over time.

Eight solid hours of sleep also increases energy, and the effect is cumulative. In Aubrey Marcus's book *Own Your Day, Own Your Life*, he presents a system where you sleep in ninety-minute increments over the course of a week rather than eight hours all at once. In any case, getting plenty of sleep is a way to help your brain, body, healing, and energy.

Finally, connecting the body directly with nature affects you on a biological scale. Make time to step outdoors often, even if it's for a quick stroll around the block. There is no reason you can't find time to breathe fresh air, stick your feet out into the grass, or sit next to a tree. If you're indoors a lot, connect with nature through houseplants. There is a wavelike state of multiple possibilities that exists in nature, and connecting with it brings us the infinite as well.

QUANTUM JUMPS AND REALITY SURFING

In quantum physics, an electron jumping from one orbit to another creates a sudden burst of energy. Usually, a quantum jump in this context is achieved by creating a large amount of energy through breathing techniques and deep meditation, where you create an idea of a bridge or a door. On the other side, there is another version of yourself with all information available to you.

While the AURA technique is about experiencing the emotion of a different reality, quantum jumping can be about accessing information. It can be used to move into parallel realities, and there are people who have integrated the two, but they are not the same.

Reality transurfing is a concept created by Vadim Zeland, and Frederick Dodson coined *parallel reality surfing*. Their concepts were about transactional, daily actions that move us into a reality using the way our minds evaluate importance and combined forces, as well as beliefs and decision-making.

Meanwhile, Burt Goldman created a detailed quantum jumping program for Mindvalley, using the concept termed by Cynthia Sue Larson. They each explore different concepts, though they all create access to other realities.

In Goldman's model, Quantum jumping is intended to be a way to access a version of yourself, for example, that's good at business—or a version of yourself who is Warren Buffett. Under this model of quantum jumping, you can access other characters as well as your own.

Napoleon Hill's *Think and Grow Rich* accessed characters that he created in his own mind. He talked to Lincoln and others who were long dead. He created these large conferences in space that he opened up in his mind, all accessed through the realities of other people who could interact with him and each other.

Quantum jumps can be powerful for writers to access characters. I have helped screenwriters move into their subconscious to interact with a character they have defined. They can be accessed for advice and guidance. I've sat back for minutes at a time as my own masterminds—which includes people like Steve Jobs and Jeff Bezos—talk through something I'm struggling with.

In Burt Goldman's theory, you are each of those people. He talks about becoming an incredible artist out of the blue, and having lived completely different existences. Dr. Joe agrees and believes that we are all one—the source is becoming united with that information. Ervin László talks about the Akasha, which is the complete history of everything, and a quantum jump would be access to that history.

Cynthia Sue Larson, a physicist at Stanford, explores the idea of actual realities shifting where people disappear completely. In this quantum jump, you are entering into a completely new physical reality different than your own. You're jumping from one universe to another in a single moment, just like an electron jumps when it moves.

To reach the full experience of a quantum jump, a huge use of energy is necessary. Before your meditation, it's important to create energy and bring it into your body and move it into your head for use in the jump. Physical movements can accomplish the former, while breathing can move it to your head.

Some people do tapping, which is tapping key parts of your body to build up energy in specific ways. Others practice Qigong exercise routines as described earlier in the book. This is why that energy routine is so important: it will give you the energy to make huge quantum jumps into brand new dimensions and realities.

Brendon Burchard recommends a couple of minutes of movements every fifty-five minutes, on a timer, throughout the day. You might not need the energy right then, but you can store the energy like batteries to be used when needed.

The specific exercise is not prescriptive—I believe you could quantum jump in *Savasana* after a really good yoga workout. If you find yourself in that sweet spot laying back at the end of a great yoga routine, it could be a good time to use that pure energy to do a quantum jump. Not everyone will want to do a rigorous ninety-minute yoga workout every time they want to jump, and it's not necessary. The key is to build up some kind of energy. Often, people find themselves falling asleep or feeling sluggish after they meditate. Quantum jumps need more energy than that.

Whatever movements or exercise you do, make sure it is with intention. Bring your body and your mind together through the movements, potentially through vocal affirmations that you say out loud.

RAISING YOUR ENERGY FOR QUANTUM JUMPING

In chapter 5, we talked about the man who visited a Tibetan temple looking for the Fountain of Youth. What he found was the Five Tibetans—five exercises that he would do twenty-one times. They include a clockwise spin, a leg lift, two kinds of stretches, and a sun salutation.[91]

After you build physical energy in these movements, spend time breathing. I practice Tony Robbin's fire breathing, integrated with Dr. Joe's breathing techniques. In this phase, you're inhaling deeply and breathing out, squeezing your abdomen and perineum as you pull the energy up and into your head.

---

91 The Five Tibetans, or the Five Rites in traditional yoga, were popularized in Peter Kelder's book *The Eye of Revelation* in 1939.

> Bring your physical energy up, then breathe to move it into your head to store it for the jump. Those ten minutes of physical energy before meditating can create the energy to make it work.

CHAPTER THIRTEEN

MIND HACKS TO CHANGE YOUR REALITY

VISUALIZATION IS NOT VISUAL

Some people struggle with the ability to visualize. There are exercises you can do to enhance the rods and cones of your eyes, but training is even more important. An important thing to remember is that visualization involves all of your senses. Rather than simply memorizing an object, visualization is trying to imagine it. When you try to create images in your mind, give them depth rather than just a visual picture.

DEVELOP SITUATIONAL AWARENESS

Wherever you go, whatever you look at, focus your full

attention on it. Look through your peripheral vision, try to become aware of signals and ideas, be open to random events happening. To help with this, imagine that this other universe deeply wants to make itself known to you and is constantly setting off little explosions all around you to tell you where to go, but like a UFO in the sky that you haven't looked up to see, the only reason you haven't noticed it is that you haven't been looking.

I think a lot of people are successful in manifesting what they want and moving through realities, but then they miss the obvious gifts that have been given to them and say it doesn't work. As you start to move into this reality, things are constantly happening around you, and you just need to be aware of it. Situational awareness training is important to complete the manifestation.

TRIGGER SUBCONSCIOUS AND CONSCIOUS STATES

Anchor states of mind which appear to be trance-like. Give yourself a cue to pull up and remember a subconscious state. I pinch my wrist whenever I am in one of these states; you might use a word, or a movement. The more you do this, the faster you will get at triggering subconscious states. Experiment with guided hypnosis and monitor your conscious states. The more you become aware of this, the more you can recreate these states. Begin to anchor different states of awareness both hyper-aware and deeply relaxed. The more time you dedicate to this the better you will get.

DEVELOP AROUSAL CONTROL

Sometimes we are manipulated by responses to outside stimuli, which can endanger us. Endorphins and cortisol pull us into pendulums, and identifying those natural responses to body stresses can help us gain control. When palms begin to sweat, hearts pound, and minds race, it's a signal that endorphins are at work. Controlling that reaction protects us in moments that would otherwise break what we're doing.

INTEGRATE SMALLER VICTORIES

Look for things that you can attribute meaning to, if there is any possible way that it's moving toward whatever reality you're manifesting. Some of these realities might take several years to move into. Don't worry about the amount of time. Instead, become aware of little goals along the way. If you were going to eat a whale, the saying goes, you would do it one bite at a time. Eventually, if you stay focused on the next bite and not the whole thing, you could get through that bad boy.

PRIME YOUR OPERATING SYSTEM

Starting your operating system the right way each morning can completely define the rest of your day. Tony Robbins does this, and it's not quite at the level of meditating—it's more like priming. I think that can become more powerful. Move into a state of deep gratitude, with your hand on your heart. Be coherent about it, bringing up incredible moments from the past—maybe with your family, the loves of your life, or your friends—and live in those moments as if they were happening right now. Pull them to you, with all of their emotions. By bringing those feelings to you, you will also begin to project them out into the electromagnetic energy that's connecting you to the source. It's priming your mind for connection.

You can change it up to experience something other than gratitude so that it doesn't become a labored task. Make sure you feel that emotion, whatever it is, even if you have to change it like you might change up an exercise. I also advocate taking it a step further. After priming, begin to act like this moment that you're in right now, no matter what it is, is the greatest moment of your whole life. How would that moment feel? What could possibly happen inside of that moment? All possibilities are available in the eternal now, even if those possibilities look like getting a flat tire or dropping a breakfast plate. Focus into the feeling of gratitude and greatness, and let the momentum build from there.[92]

---

[92] Try out my twelve-minute morning priming routine on episode 45 of *The Reality Revolution Podcast*. http://www.therealityrevolution.com/the-12-minute-morning-priming-meditation-ep-45/.

USE EYE MOVEMENT TO ACCESS CONSCIOUSNESS

Imagine that your eyes are like a joystick for your consciousness. You can access different sense modalities and create them as well. Neurolinguistic programming analyzes the way people move their eyes to access different states of consciousness. The left side of the brain controls the right side of the body and is typically responsible for tasks that have to do with logic, such as science and mathematics. The right side of the brain controls the left side of the body, as well as creativity, beauty, and art. It's different depending on whether you're left-handed or right-handed, and both sides certainly work together. But if someone looks at you and says a lie, they are going to look up and to the creative part of their brain.

Then, there is the eidetic memory, which is what you already have, and it's broken up into visual (up), auditory (to the left), and feeling (down). People who are very kinesthetic look down a lot and have a certain body language and tone of voice. They are filtering the world through their bodies. People who are auditory will use sounds and patterns in their communication and will remember by looking to their left.

The first hack to this is to take your eyes through all six parts of your body, as well as to a seventh part: the nose. If you look directly down and directly up, you access sense of smell and taste as well. Spending fifteen seconds looking each way is said to open the third eye—others believe the opening of the pineal gland is connected to rapid eye movement.

EYE MOVEMENTS

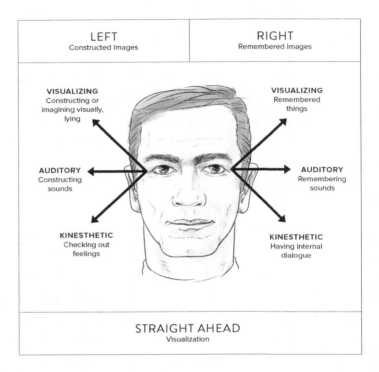

In order, that's looking left-up, left-down, right-up, left-up, left-down. Then right-down, right-up, straight-down, straight-up, straight over to the left, and straight over to the right, each for fifteen seconds.

If done regularly, eventually this mind hack can quicken to one second each, like you've become fully aware and able to access all of the compartments of your mind. This is excellent for creative work, becoming an anchor that can open up more neural structures and a higher level of awareness in your brain.[93]

FORM A DAILY CREATIVE PRACTICE

Next to my bed, there is a five-foot oil painting of elaborate, infinite universes on a rocket ship. I made it my goal to be creative every single day, and that painting is both my creative work and my reminder to continue. Look around your house and tell me, what did you create?

While the realities that we're maneuvering through do exist, we use the creative parts of our minds to access them. Neville Goddard called it imagination. Some people become blocked because they believe they "aren't creative." I don't believe that's true. A rancher who has done nothing more than take care of cattle every single day might not have done much art in his life, but that doesn't mean he can't be creative. This hack is to simply do something creative every single day. This can be as simple as writing a brand new sentence or drawing, a squiggle, an idea, a new recipe. One of my favorite things is to make a playlist. If you can do something more elaborate, do it.

Once you make this a practice, you'll find yourself becoming more and more creative, until you start creat-

---

[93] Try the advanced holographic visualization technique for yourself on episode 57 of *The Reality Revolution Podcast*. http://www.therealityrevolution.com/the-advanced-holographic-visualization-technique/.

ing out of desire and not simply habit. The creative part of your mind relaxes and begins to access the space of variations in its own way. This will give you more and more power to access better and better realities that the creative part of you had limited before. You become freer to enter into the unknown.

USE YOUR NONDOMINANT HAND

This is a tidbit I picked up from a book called *Thought Revolution* by William Donius: if you write out a question with your left hand when you're right-handed, you're accessing the other side of your brain to get the answer. When he asked questions and wrote out the answers with his nondominant hand, he started to come to incredible answers.

The first few times, this was scary for me. I couldn't read what I was writing. Eventually, this simple exercise can unlock deep insights.

LAUGHTER REALLY IS MEDICINE

When the body is sick, it thinks that it is the mind and wants to pull all of the focus onto the illness. If you can laugh, especially for periods of time, you can break out of that spiral. Laughing at a movie or a TV show or a book is spiritual work. It's a brain hack that diverts your path toward positivity that can help you ride the right waves.

CREATE PERMANENCE IN A JOURNAL

There is a permanent reality that occurs when you start to journal. Every single self-help guru will tell you to journal, and that the best way to do it is with gratitude and goals. Few of them talk about asking yourself questions in the journal. I recommend writing yourself questions before bed, then answering them in your dreams. When you wake up, write down what came to you.

The more you write down, whatever you write, the more you process. I have fifty journals spread throughout the house. I rarely go back to read them, but the process of putting pen to paper is transformative.

SET MORNING AND EVENING ROUTINES

Priming and journaling are excellent morning routines, but there are more patterns to lock in to as well. It's absolutely critical to settle into a routine in order to get your mind in the right place.

ESTABLISH CONTROLLED DECISION-MAKING

Making decisions is an energy, which is why Steve Jobs famously limited his decisions. Every moment contains millions of options to choose from, and every time you make a decision, you're not only using up gasoline in your energy tank, but also creating more realities. Accumulated options will, over the years, shape your reality, and the conscious and subconscious choices can bring us into better realities where the options become better as well. If you can begin to control where you make decisions in the smaller moments, such as with the clothes you wear every day, you can bring more energy to the other moments of your life.

HYPNOSIS

Hypnosis is an old study of the way language works to communicate to the subconscious. A hypnotic trance is not therapeutic in and of itself, though specific suggestions and images fed to people in a trance can profoundly alter their behavior. Hypnosis allows them to rehearse new ways that they want to think and feel, and to lay the groundwork for powerful changes in their future actions. For example, people who use hypnosis to quit smoking will go hours without thinking about a cigarette, then when they do light one up it will taste terrible.

I've talked people through the imagery of being a nonsmoker, of finding that they breathe easier, having more energy, enjoying subtle tastes and smells more, and feeling good about their health. Those images deep within their minds start to change their outward behavior.

The deep relaxation of a hypnotic trance is also broadly beneficial, as many psychological and physical illnesses are aggravated by anxiety and muscle tension. A growing number of studies, adding to the body of research indicating safety and effectiveness, show the benefits hypnosis might have for childbirth pain, headaches, concentration and study habits, the relief of minor phobias, and anesthesia.[94]

We might think that we can't change or that we're stuck, but it's our subconsciousness pulling us along and creating that perception. Hypnosis is one way to communicate directly with the subconscious and change those patterns.[95]

REPRESENTATIONAL SYSTEMS AND SUBMODALITIES

Neurolinguistic programming opened up the idea that we understand the world through our own representational systems, and that we become biased toward those systems. For some people those are rooted in visual things, for others it's in sound or taste and smell. The study of NLP uses language to describe and model human experiences in the way that we filter information.

Within the study of representational systems and NLP, the concept of submodalities arose. This is taking the systems beyond the five senses and into deeper layers. For example, we could describe a picture as being "black and white" or "color," or using submodalities we could describe it as being bright or dim. Sounds could be loud

---

[94] More on this in *Hypnosis Heals*.

[95] Check out https://therealityrevolution.com to access over 100 different hypnosis programs.

or soft as well as coming from a certain direction. Feelings could be in different parts of the body as well as with different temperatures. Smells could be pleasant or offensive, strong or light. Taste could be sweet or bitter, strong or mild. These finer distinctions are submodalities, and they define the qualities of our internal representations. I have found this to be an incredibly powerful way to code internal experiences and ultimately to maneuver realities.

Research indicates that the brain uses these structural elements as a way to know how it feels about things and what they signify internally. Generally, we work with the modalities of visual, auditory, and kinesthetic. Within each, there are layers of submodalities—for example, when people's eyes quiver—that indicates an internal feeling, because our feelings are attached to functions in the body. The link is understood as bilateral, meaning emotions attached to mental experiences are affected by submodalities and specific submodalities can be affected if the emotional significance changes. Discovering this bilateral influence—that emotions affect thoughts, and vice versa—and focusing on them both visually and meditatively has led to huge transformations for people.

Visual: size, distance, brightness, color, frame, two-dimensions or three-dimensions, clear, fuzzy, shape, movement, still photos, slideshows, moving, looping

Auditory: mono, stereo, volume, pitch, tempo, rhythm, looping,

> fading in, voice, background sounds
>
> Kinesthetic: somatic sensation, location, movement, traction, pressure, texture, temperature

MODELING

The idea is to look to people who are incredibly successful and model their behaviors. They have specific daily habits, body movements, breathing techniques that create the reality they are in. They are in a particular vibration for everything they do, from the time they wake up to the way they breathe and the way they move their eyes. Body mimicry can bring us into similar vibrations.

Most people in the self-help community got the concept of modeling from Tony Robbins, and he talks about teaching people in the military how to shoot guns. The interesting part is that he had never shot a gun himself. But he identified the best marksmen and broke their behaviors down into a model. The efficiency of the marksmen in his program increased by 75 percent.

If people have achieved a reality similar to what you want, go read a book that they've written. Listen to them speak. If you have a chance to be mentored by them, it's even better. Get into their minds and habits. You're not imitating them, because you're still uniquely yourself. Modeling isn't about losing who you are, but about realizing that your identity is more flexible and expandable than you assume. Your core remains intact while you expand your identity to model the reality you want.

SWISH PATTERN

This powerful technique came from early linguistic programming and hypnosis, and it works well if you have a habit you want to overcome or a behavior you want to adapt. Quitting smoking is the most common instance, where we would associate smoking with something unrelated and undesirable. Then we create a visual picture of both and switch them—integrate them together in your mind.

When you do this switch patterning over and over, you start to associate feelings for the cigarettes or whatever habit you want to overcome with something disgusting or terrible. Another way to approach this is with associating something positive but difficult, like getting up earlier in the morning. Maybe you feel grumpy about it right now, but if you switch patterns with incredible vibrancy, joy, happiness, and energy, you can create an association that you might not have had otherwise. That kind of ingrained association can bring your normal life in line with new patterns and programs.

REVERSE VISUALIZATION

The brain learns things quickly, not slowly. If you were

to watch one scene from a movie each day for five days, you'd never get to the plot of the movie. We need rapid sequencing of frames to get an idea. Reverse visualization is a technique that works best moving quickly.

When we're thinking about moving to a parallel reality, we're trying to move into something better or to overcome some problem or issue—often a phobia or an addiction. Imagine you're in a big theater, sitting near the middle of the room. On the screen is a still photo in black and white, in which you see yourself just before indulging in the fear or activity that you want to overcome. Reducing the submodality of color and sound can help you lock the image into your brain as you visualize that moment.

Now, imagine floating out of your body from your seat in the theater up to the projection booth. From the booth, you can still see yourself sitting in the seats, clearly enough that you notice the color of the shirt you're wearing. Now, turn that still snapshot on the screen into a black-and-white film. Watch it from the beginning to just beyond the end of the unpleasant experience, then stop it. Jump into the projector and physically run the film backwards, like you're rewinding the movie.

This isn't an easy exercise, but when you can do it, you'll watch the scene reverse itself. Something about that reversal creates a physical change, undoing that pattern in other parts of your brain as well.

GRATITUDE

This is a topic that continues to come up because it is the emotional signature of the reality that you want. Express gratitude for something you haven't received yet. Understand what that feels like—you have achieved some incredible transformation and have cultivated a feeling of gratitude around it, even before it happens. That emotional signature will bring on future events so that you will be more grateful.

Psychologists Robert Emmons and Michael McCullough explored what might happen to people's happiness levels if they were reminded of good things that were constantly present in their lives.[96] So three groups of people were asked to spend a few minutes each week writing. The first group listed five things for which they were grateful, the second noted five things that annoyed them, and the third jotted down five events that had taken place. Not surprisingly, the group that was grateful ended up more optimistic and happier about the future. They were physically healthier and even exercised more than the others.

The feeling is the secret. Gratitude is more than just a trick that makes you feel good. It's broadcast out as an emotional signature, and it can attract reality that includes the thing you're grateful for.[97]

---

[96] Robert Emmons and Michael McCullough, "Counting Blessings Versus Burdens: An Experimental Investigation of Gratitude and Subjective Well-Being in Daily Life," Journal of Personality and Social Psychology, Vol. 84, No. 2, (2003) 377-389.

[97] Try my gratitude meditation: http://www.therealityrevolution.com/the-gratitude-meditation-111hz-396hz-372-hz-ep-86/.

CONCLUSION

Deep inside of us, there is a symphony of a hundred billion neurons firing in concert, constructing the vivid reality in which we live. We are inextricably connected to our minds, bodies, and consciousness. After years of research, practice, and experimentation, this book is the culmination of patterns and techniques to connect you to that power. I want to light the spark and watch you make your dreams come true.

Anyone can make millions of dollars. What can you really do in your life? You are your own genie in the lamp, if you want to do the work.

For many of us, those promises seem like a dream. Can we really fulfill our own wishes? If you're reading this book, it's possible that you've tried to before and haven't been able to. If you're hanging on to doubts about your skills, feelings of unworthiness, fear of failure, or lack of confidence, know this: it is not too late.

You are not your past.

Whatever you think you did or where you believe you failed, it doesn't matter. You're here now. You're in this moment with me today. You've attracted this book to yourself for a reason, and the knowledge within it can transform your life, this time for good.

I know what it is to be overwhelmed with fear, guilt, and regret. I know from experiences that entertaining those beliefs and thought patterns become self-fulfilling. You won't be able to reach your goals. You won't be able to have the relationship you desire or the job you want or the business growth that you need. I want to offer you a way out—an alternative mindset filled with love, gratitude, and hope.

LOOKING TO THE FUTURE

This world is a crazy place right now. People are being exploited for low wages, with the gaps between rich and poor growing all the time. There's ecological collapse, and a movement of hatred that seems to be filled with wars, famine, disease, and politics that tend to pull us toward our fears. At the same time, we are absolutely facing a quantum age in which the power of our minds to create reality is increasing. Everything seems more intense, perhaps because we are becoming more adept at creating our own realities, whether or not we know it.

The power of our thoughts is going to increase over time. Technological advances in every major field, from computers to pharmaceuticals to DNA and CRISPR indicate a world in which we live incredibly long lives or create materials out of pure information. Imagine that. We could sit down to write at a computer and the computer helps to distill your thoughts. We will be able to look into the past to solve problems that right now seem insurmountable.

Imagine a future where the realm of possibilities is entangled and integrated into the one we live in.

There is no denying that the future is going to be different than anything we can imagine in this moment—the only question is how much intention we will bring into that transformation. The more adept we all become at understanding quantum physics and parallel realities, the more we will shape the world.

It's exciting to think about a future reality filled with purpose, hope, abundance, and happiness. My sole hope for this book is that you will begin to believe that anything is possible. That the mind works in unique ways to give you what you want. That we truly can overcome the global warming, trade wars, disparity, and things that we don't currently believe we can solve. That the human mind has remarkable faculties for empathy, love, compassion, and solidarity.

Creativity and love are not genetic traits given to some people but cultural endowments to all human beings. This creative power can create a reality revolution not unlike the industrial revolution of the past.

We can choose in each moment to bring love or hate

into our own lives, which collectively brings love and hate into the world. This is the reality revolution that will transform the world.

AFTER THIS BOOK

There are multiple initiatives that you can participate in that focus on global consciousness. The Club of Budapest and the Global Consciousness Project focus on changing global consciousness to create better tolerance, understanding, peace, and joy with all of our neighbors. Back in 2007, we held a global peace, meditation, and prayer day, where sixty-five countries prayed for peace at the same time. The more we focus as a group, the more we can change the world.

Try the exercises throughout the book, working especially carefully on finding true love for yourself. Embrace the idea of unlimited possibilities. Use the resources listed to further the techniques you've developed in this book. There are a number of mediations, strategies, and coaching available to you that will help you explore parallel realities, increase your health, and transcend whatever limitations that you have.

This is not an easy undertaking. I've had the fortune of studying many books and meeting with many people who have experienced radical transformations in their lives. You cannot turn a basic belief and a few minutes of meditation into a manifestation. It just doesn't work like that. What I've compiled in this book is a thorough understanding of the quantum age—the self-help of years past updated for a new generation. These are steps that I follow, and my dreams have come true.

As I open myself up to possibilities and spread the love and energy out from what I've learned, it has all come back to me tenfold. My transformation from caterpillar into butterfly has taken me beyond anything I could have imagined. Every part of me wants you to experience this transformation as well.

We're about to hit a collective tipping point. This is the moment we are in right now, about to experience a revolution. We can choose what that looks like. We can fill it with love, happiness, and joy. Thank you for exploring this potential with me. Thank you for stretching yourself beyond your understanding of the universe. Thank you for believing in the power of your own mind.

Welcome to the Reality Revolution.

> Life is short. You are not promised anything tomorrow or next month or next year. If it were to all end tomorrow, have you lived the life that you wanted? Instead of settling, consider this a message from your future self: When your heart stops, don't let that be a mere formality. Start each day knowing that it will all be over eventually.
>
> The future is incredible, but your days are numbered. Don't fear death—cherish life. Death is the best invention of life. We're

all going to die, so there's no reason not to follow your heart. To go after your dreams, and to do it with passion. Choose kindness over being right. Choose the hard task instead of the easy one. Release your fears of rejection. Stand up more. Love more. Live your legacy more.

Aim to be extraordinary. *You are extraordinary.* You are infinite. Go the extra mile. One more rep. One more second. One more stretch outside of your comfort zone. Be willing to fail. Get rid of negative people in your life, but be the person that others can depend on. Depend on yourself. Be known by your actions. Wake up and choose to become kindhearted, and to experience kindheartedness. Stop complaining and start loving.

Stay uncomfortable. Do the spontaneous thing. Do the things you're afraid of. Don't just think positively, but actually become the embodiment of the person you want to be.

If you want to harness the power of the universe, you must be willing to change your thoughts. To change your ways of being. You're a perfect reflection of the choices that you've made, so make better choices. Believe in miracles. Believe in kindness. Believe that great people are coming into your life, there's nothing you cannot do, you cannot have, you cannot be. Don't believe in luck—believe in the power of your thoughts. Believe anything is possible.

Because it is.

ACKNOWLEDGMENTS

This book would not have been possible without:

Nikki Van Noy, who helped take my pile of words and turn them into actual sentences.

Kacy Wren, who helped keep this book on schedule.

Hunter Bengtson, for telling me, "Dad, don't let your dreams just be dreams!"

Hayden Culver, for teaching me about life.

Nancy Mena, for helping me to see what is possible.

Gene Bengtson, for showing me the joy of reading.

Cheryl Bengtson, for showing me the joy of teaching.

Tom Wright, who taught me the joys of telling a story.

Wendy Benning Swanson, for inspiring me.

Darwin Summer, for answering those late-night phone calls of despair.

Frank De La Cerna, for being there for me when no one else was.

Cynthia Sue Larson, Frederick Dodson, Aaron Abke,

Owen Hunt, Sunny Sharma, Quazi Johir, and Lawrence Watusi for teaching me about the universe.

And to all of those listeners and viewers of *The Reality Revolution Podcast*, for sharing my explorations of consciousness and reality.

ABOUT THE AUTHOR

BRIAN SCOTT is a proud father, writer, painter, entrepreneur, epiphany addict, inner space astronaut, life coach, transformation engineer, futurist, awe junkie, wow archaeologist, quantum jumper, serial entrepreneur, hypnotist, neurolinguistic programmer, meditation instructor, motivational speaker, researcher, intuition teacher, mixtape connoisseur, luck instructor, survivor, quantum success artist, Pearl Jam lifer, future owner of the Denver Broncos, and founder of the Advanced Success Institute.

The Reality Revolution is born out of a fanatical vision quest to understand a near-death experience which was the culmination of a profound spiritual awakening in which Brian explores whether he has shifted into a parallel reality. The mission of the Advanced Success Institute is to explore the new movement to hack reality and transform lives by exploring experiential quantum physics, reality transurfing, quantum jumping, meditation, hypnosis, qigong, sensory deprivation, virtual reality, mind tech, ayahuasca, psychedelics, channeling, manifestation, mindfulness, neurolinguistic programming, epigenetics, EFT, energy psychology, yoga, luck coaching, and so much more.

Made in the USA
Las Vegas, NV
21 December 2023

83421013R00236